Time to Tell:

a look at how we tick

Time to Tell:

a look at how we tick

Ronald Green

BOOKS

Winchester, UK
Washington, USA

First published by Iff Books, 2018
Iff Books is an imprint of John Hunt Publishing Ltd., No. 3 East Street, Alresford, Hampshire SO24 9EE, UK
office1@jhpbooks.net
www.johnhuntpublishing.com
www.iff-books.com

For distributor details and how to order please visit the 'Ordering' section on our website.

ISBN: 978 1 78535 695 7
978 1 78535 696 4 (ebook)
Library of Congress Control Number: 2017933414

A CIP catalogue record for this book is available from the British Library.

Design: Stuart Davies

Printed and bound by CPI Group (UK) Ltd, Croydon, CR0 4YY, UK

Contents

This book is dedicated to a woman I saw working in a brick factory on the Mekong Delta, Vietnam. She works nine hours a day, seven days a week, by a furnace, scraping ashes from the sides, scooping up the hot ash from the floor and depositing the heavy load on another pile. Backward and forward all day. Looking at that emaciated woman, who may have been in her early thirties, although it was difficult to tell, made me realize that the concept of 'rest,' never mind the notion of reading anything in her 'free time,' would be completely unknown to her. There is no free time for her. Time simply isn't free. If time is anything for her, it is survival.

Seeing her gave me a new respect for time within the awareness of the privilege I have that she doesn't. It is the privilege of having time to write this book.

I know not her name, nor will she know of this book. But it is to her that I dedicate it.

In loving memory of my mother, Etta Green, 1922–2015

Acknowledgements

With time being such an all-encompassing topic, it isn't surprising that so many people were willing to discuss its aspects, give their advice and argue with me. Each one of them expressed thoughts and points of view that helped me weave together the strands of my own perspective and form my own theory of time. I thank each and every one of them.

In particular, I thank those without whom this book would not have appeared:

– Debbie Knoller, whose span of philosophical knowledge and insights were invaluable throughout the development of the book. Our discussions were long, and her patience was legion. Often we had to agree to disagree. But it is due to her encouragement that I continued when time threatened to overwhelm me.

– Frank Shapiro, whose perspective and knowledge of history were inestimable, and whose back and forth reading of the historical parts of the book tremendously improved the book as a whole. Although we did not always agree on all aspects of history within time (and vice versa), his encouragement and friendship moved me on.

– Roland Mann, whose insights, breadth of knowledge and ideas uncannily reflect my own worldview. Our many, many hours of talks sharpened my views and broadened my outlook, allowing me to wander into aspects that I had not previously considered. To him I owe, and am thankful for, an invaluable friendship.

– My wife, Lily, whose infinite patience was the rock on which I could survive the mangled path through time and emerge relatively sane. The countless hours per day I disappeared in my pursuit of time were taken in stride by her and I can but hope that I will make up for them now that that time is behind me.

Although he didn't realize it, it was my online discussions with Gunther Lippert that caused me to choose the subject of 'now' as central to my book about time. For this I thank him. Our sometimes heated deliberations brought out the centrality of the notion of 'now' within how the world is widely viewed, and the subsequent discussions that 'now' engendered brought home to me how deeply people are affected by 'now.' My thanks go also to the many participants in The Virtual Muser forum on Facebook, who expressed their views and whose lively views brought the topic to life.

It goes without saying that any mistakes are mine and nobody, apart from me, is responsible for the conclusions and views expressed in this book.

Preface

We are constant time travelers, all of us. Always, every second, whether awake or asleep. We don't actually do anything to travel in time. We can't help it; it happens, whether we want to or not. Time passes, and we with it toward the future, never getting there but always on the way. It's quite exhausting if you think about it. But mainly it is—or should be—exhilarating, this continuing movement through what we didn't know before to what we will know just around the corner.

So what, why, how, when?

Everyone has something to say about time. Obviously, because time is the heart of nature, and it is the heart of our own personal nature. There is no consciousness without time. Anything and everything can be asked about time, because all answers will shine a light upon us as humans, about our behavior, about who we are and why we are. It was the Russian philosopher Alexander Herzen, when referring to the expression 'time is money,' who said that, *"In reality, time is much, much more precious than money:* time is ourselves."

This book, then, is about what time is for us, what it does to us and for us, and how we live and react to it.

It is not plain sailing, though. How can it be, when our journey into time shakes up our comfort zone by questioning what we naturally take for granted?

And where can we start if not from where we are—the present? It is our *when* and *where,* the moment in which things seem to happen, the point from which we look backward to the past and forward to the future. As such, the book will shine a particularly bright light on the concept of ***now***, taken for granted as the central peg of life and living. It is here that the rumblings of our earthquake begins.

Every age has its deities. The Western medieval mindset

placed its blind faith in God. The Enlightenment anointed reason and science. Our own age has indulged a cult of getting things NOW, which is a seismic shift from the idea of future consequences to that of immediate satisfaction. We can see this starkly in New Age and Eastern faiths with their ***now***, in which everything occurs because ***now*** is supposedly all there is.

It is not so much what scientists tell us about time that this book is about; much has been written about that, and physicists and cosmologists will be kept busy for a long while delving into the properties of time.

Not surprisingly, nobody has yet succeeded in defining time, all the way back to Augustine, who said: "If no one asks me, I know. If I wish to explain it to one that asketh, I know not." Umpteen generations later, French philosopher Henri Bergson answered with a quip that "Time is nature's way of preventing everything from happening at once," which is pithy and not untrue.

And just when we thought we knew something about time, it took Einstein to show that he didn't have a concrete answer either, since his theory was that time is relative depending on where you are, and it is affected by whatever does the measuring. The most commonly accepted definition, endorsed by Einstein, likely with a twinkle in his eye, is: "Time is what clocks measure." Now while his theories of time are infinitely more complicated than that in their explanation of how and why time is relative (and why, in fact, you couldn't even say with authority what time it is), we all go along with the connection between time and clocks. But it's much, much more than measuring time, since whatever we do has a connection to time in some way or another. We cannot, in fact, grasp anything that is not within time and space.

What is important, then, is what time does for us—in other words, not what time *is*, but what it *does*. There is, in other words, no definition of time that does not take into consideration the place of humans within it.

It is when we question this very fundamental of where we are—the present—that the next step in our understanding of time is taken. The repercussions of such a step are breathtaking.

One of the most fascinating things about time is that once we begin thinking about it, we stop to look at things we normally take for granted, like why time seems to go faster when we are enjoying ourselves, and why it seems to slow down to a miserable crawl when we are somewhere we'd prefer not to be.

Thinking about time can also lead us to places we had perhaps fleetingly visited but at which we rarely stop. So there are the bigger questions, such as: wondering about what is real and what perhaps isn't; whether how much of our memories are what actually happened; wondering if things we are sure about are true or not, or wondering what is our imagination and what is someone else's imagination.

Time really is weird once you begin to think about it. And thinking about it is what we will do. Not just thinking about it, though. Thinking leads us to challenge our own point of view, to rattle that cage and act upon what we discover.

If in my previous book, *Nothing Matters: a book about nothing,* I dealt with what isn't, in the book you are now taking time to read I delve into what is potentially everything. If *Nothing* is timeless—in fact, it is only *Nothing* that has no time—everything else has time: where there is anything there must be time.

This book, light on descriptions in physics or cosmology, does not explain time through quantum theory, relativity and speculative gravitational geometries, even though now and again I may have to glancingly touch on some of those, and show, surprisingly, how often our observations reflect those of cutting-edge science. It is a book of thought and of thoughts that will be triggered by some of the things we look at. If it happens that now and then our thoughts grind against certain modern theories of science, we need be only a little worried about that;

for every scientific theory, there is another one that challenges and disputes it. Interesting, though, will be the fresh slants on science that will actually enhance and give human explanations to dry theory.

It is possible that some of the conclusions in this book will undermine popular and common basic beliefs of how things are and how they work. It is more than likely that I will tread on some toes when I look at the notion of 'now,' the landscape of memories and vistas of the future. It follows, then, that not everyone will agree with them. But we don't expect everyone to agree with everything in any case; so with something as nebulous as time I will be doing well if people are stimulated to look anew at different aspects of time that we confront daily.

So we are going to keep our feet firmly on the ground (even if it shakes) and if we fly into space, or theorize about traveling at the speed of light, we will do so as thinking beings. Any time travel we will be doing will be in our minds, back into our memories and forward into how we potentialize what will come.

A question that needs to be asked is why knowing about time is so important and why giving a new slant to time is even more so. In normal circumstances we need to know what time it is at the moment, so that we can plan our activities around what is and what is planned to be. What difference does it make, for example, what 'at the moment' is exactly, or whether our memories are of what really happened? Perhaps on a day-to-day basis, when we go about our normal lives, it doesn't. But when we do think about them, the chances are that these questions, and others that lead on from them, make up the key to a greater understanding of where we are within our surroundings and why we react the way we do.

Time is everywhere and is everything. Time is relative; time is money; time is space; time is being; time is spent and it is lost; time is long and it is short; we pass time, we take our time, we finish in the nick of time and things take time. Time is all of those things, and we will be looking at them—and more—from a fresh perspective.

We are, after all, a constantly moving point at the end of where we have come from or, as the philosopher Thomas Nagel puts it: "we are products of the long history of the universe since the Big Bang, descended from bacteria through millions of years of natural selection."

The fact that time dominates experience, presented me with an immediate practical problem when writing this book. My initial idea to divide the book into three sections, the present, the past and the future, broke down with the realization that there are no real boundaries between these three 'times.' There are, as is obvious, not 'times'; there is only 'time,' a seamless interwoven phenomenon that reflects what is happening and why—in fact, what we are and why.

And let's not get carried away by the connection of clocks with time. As important as the human invention of dividing up the passage of time into units is, it is just a way of us trying to put some order in the changes that occur around and inside us. But time is change and since everything changes, time exists whether we measure it or not.

There is a lot to be said about time, and how we deal with it and it with us. Our feelings about time are very strong, since everything we think about is connected, including beginnings and endings, life and death. Time, the boundary between being and not being, allows us our freedom and shows us our limitations.

It's no wonder that time has always been a central discussion point in disciplines throughout history and continues to be so today. How we deal with time is a reflection of how we deal with ourselves and the world in which we live. The urge to find out more about time is the urge to find out more about what makes us, us. It never hurts to look again, in a new way. When that allows us to forge a way forward that is best for us, it will have been worth it.

I

Is

Happening together

It's all about time. Everything is. Yet, of all the strange things we take for granted, time must be the strangest. And we don't need to scratch very far to discover that what we experience is not necessarily what is happening.

It goes without saying, for example, that things are happening now. But when is *now*? As crazy as the question seems at first—after all, *now* is, surely, right at this moment in the present—the answer is not at all simple and self-understood the moment we think about it. It's somewhat of a trick question, though, since any instant of *now* will be over long before you have read this far.

Now is not easy to land on or even define, even though we're sure we know when it occurs. After all, we have no difficulty in distinguishing it from the past and the future. We use it constantly in our day-to-day language and don't think twice about it. You are reading these words now, aren't you?

Yet, in which way is it 'now'? Surely 'now' is lasting too long to actually be *now*. Frustratingly, although we are aware roughly when *now* is, we can't catch hold of it before it is gone.

So is there any way we can know when *now* is? Is it ever *now*? Or is it always *now*, as many believe?

So what? you may ask. Why does it matter? If it is our strong intuition that what we do is taking place now, and that it is a period of time that we call the present, why should we question it? What's the point?

It does matter, as we will see. If *now* is where we think we are, how will it affect the way we live our lives if we find out that *what* we are depends on *when* we are? And while intuitions are important, they may not represent what may be actually happening. Intuitions are not 'right' or 'wrong,' though; they are feelings, and feelings are real. But however genuine they are, feelings lead us down paths on which we might not necessarily want to be, causing us to behave in ways we'd prefer not to.

So let's pause from time to time and look at what we so readily take for granted. We can choose between the blue pill, continuing to go along with what we take for granted. Or we can lean over for the red pill. Who knows if the red pill will give us definitive answers—'the truth of reality' mentioned in the movie *The Matrix*? And why even assume that there is 'the truth' out there somewhere? Perhaps, after all, we'll discover that our intuitions were correct all along. Or perhaps not. In any case, let's throw aside the self-help and how-to-live books while we think for ourselves, and, as an extra, the opportunity to think about how we think.

It goes without saying that we are unable to do everything at the same time. Actually, it's impossible to do, or think of, even two things at once!

Pretty shocking that is at first glance, especially as 'multitasking' has become the apparent answer to the ever-increasing demands of time in the runaway 21st century. After all, if we are able to read a book while listening to music, or have no difficulty watching a movie on the television while at the same time doing the ironing, it does look as though we can do all sorts of things at the same time. Yet we can't.

With each of those events, some of the tasks are so automatic that we don't actually think about them at all. So while it is true that when driving we can listen to a talk on the radio, our mind will switch off the radio if something happens on the road that needs our attention.

It's with the non-routine tasks that we are frustratingly stuck at being able to do only one at a time. Despite all the talk about multitasking, sadly there is no such thing.

This became very clear to me while writing this book you are now reading. Thinking I would work faster by having two or three websites on different topics open simultaneously on my computer, while also having open on my desk a couple of books that I was consulting, it didn't help. Now while the sites were 'open,' it was still necessary for me to click on each in order to bring it up on my screen, or move my eyes from one to another on a split screen in the same way that I had to move my eyes from one book to another.

My own case showed that the effort was even slowing me down, since I was jumping from one task to another in a sort of hyperlinked fashion, sometimes losing track of where I had started and where I was going.

Not only me, of course. Ulrich Mayr, a professor of psychology at the University of Oregon, who studies multitasking, made that very point at a conference in Stamford, Connecticut, in 2009, when he emphasized that "multitasking is actually rapid task switching, since the human brain does just one thing at a time."

A fascinating demonstration of this is on a YouTube video, where viewers are asked to count how many times the basketball players in white shirts pass the ball. The answer was 16, but that wasn't important. Almost half the viewers were so focused on the players that they did not notice a person in a gorilla outfit who walked through in the middle! When told about this, they watched the video again, this time noticing the gorilla and wondering how they hadn't seen it the first time round. But what they missed this time also was that the background curtain changed color and that one of the players in black left the picture.[1]

When looking at the gorilla, we are not looking at anything else. We cannot look at anything else, since we can only focus on one object at a time, with light reflected from each hitting our retina in turn. When we move our eyes to another image, we receive the light from that image and cease getting it from the previous one. It is rapidly moving our eyes from one image to another that gives us the impression of seeing a number of things at the same time.

It happens to us often that we notice something only when it is pointed out to us. Imagine watching a football match being screened on TV as the camera pans past all the spectators in the front row. When later, your friend who had been sitting in that front row asks if you had seen him, the answer would probably be that you hadn't. If you then watched the scene again, you would see him because you were looking for him. There is much that flashes past us, a lot of which we miss for the simple reason that we can see only one thing at a time. This is, of course, how magicians do their tricks, getting us to focus on something they want us to while adjustments are made elsewhere away from our focus of attention.

That presents us with an interesting question: Does the fact that we cannot *do* more than one thing at a time also indicate things about the world, namely that things cannot *happen* at the same time?

If two stones simultaneously hit a wall near us, we hear only one blow as they hit; and if we videoed the event, we would likely recapture both stones seemingly hitting the wall together.

But do they? Fortunately, we can test it. If we timed it using an accurate enough timing device, we'd discover that one stone hit the wall fractionally before the other. It comes down to how we time it, of course. If we timed it in full seconds, we wouldn't know about their different times of impact, since the stones would hit 'at the same time,' that is, within the same second. If, however, we timed it to, say, a tenth of a second, we'd readily note the stones hitting at different times.

Why does this matter? Mostly it doesn't, since we aren't usually concerned about things being so exact. If our train is scheduled to depart at 09:05, it wouldn't bother anyone if it does so at 09:058 or even at 09:07. And if, while on the train waiting for it to depart, we phoned someone we had sent off to buy a drink and a sandwich, to inform them that they had better hurry because the train was leaving 'now,' we would mean that its departure was imminent, not necessarily that the train had started moving at the moment we uttered the word 'now.'

There are, though, many practical situations, in which it is essential to know exactly when something happens and what event came before another one. In competitive sports events such as athletics and swimming, until fairly recently it was possible to have tied results; but they were tied simply because we didn't have a method to time accurately enough which competitor reached the finishing line before another. Close results used to be decided by officials with stopwatches, a notoriously fickle and incredibly inaccurate way of deciding who came in when. Today, races are measured electronically to a hundredth of a second. Take as an example the win of the 20-year-old South African swimmer, Chad le Clos, over US super medal holder, Michael Phelps, by a mere 0.05 seconds in the finish of the 4 x 200m freestyle relay at the 2012 London Olympic Games. Obviously this win was not immediately apparent to the spectators, for whom both finished at exactly the same time. In fact, had the race been timed by a clock that measured time in full seconds, or even in tenths of seconds, the spectators would have been right: there would have been two winners. Only because we have the means for finely calibrated times, can we see what the real situation is. It's a relief, of course, since we can safely say that sports events with time as the deciding element will never again be tied.

And it does make us think. For a start, it makes us wonder if it is possible at all for *anything* to arrive at the same time as something else. On a day-to-day experience, they seem to—but only because our eyes and brains aren't capable of timing things to within a tenth or a hundredth of a second. But if we were able to do that, we would always see what arrived before something else, so seeing events in slow motion in a way similar to a seagull's ability to spot a fish in the sea and swoop down to catch it. It's why flies avoid being swatted by our rolled-up newspaper; for them, the newspaper that we swing at lightning speed is inching toward them extremely slowly.

Even though we personally, without the help of technology, are unable to discern such tiny differences in time, it seems to be irrefutable that an event cannot *finish* at exactly the same time as another event, and that it cannot do so because they can't *happen* at the same time. It follows, then, that no two things can *begin* at the same time, either.

It doesn't matter how accurate the timing device—even an atomic clock that can measure increments of time out to 17 decimal places—there is, in fact, no such notion as 'the same time,' since, according to Einstein's special theory of relativity, it depends on the location of the events being timed. It is impossible in an *absolute* sense that two distinct events occur at the same time if those events are separated in space.

That would leave us with a puzzle. If we can't time events accurately, when exactly do things happen? And when we talk about something happening *now*, when is that?

If you watch our stones hit the wall, how could you decide which one of those stones is hitting the wall *now*? If they both don't land at the same time—and we know that they don't—one of them came before the other. Does the first one hit *now*, or does the second? Do either?

So when exactly is this *now* that we all feel so intuitively to be real as to not think about it? Perhaps *the present moment* is clearer? No, not really. It is as problematic as *now*. Let's face it: *the present* is itself an amorphous lengthy period with fuzzy edges; we don't really know where it begins or where it ends, when it is here or when it has already become then.

In our daily lives, none of this bothers most of us. Why should it? We seem to live well with the term *now* and we are comfortable in *the present*. After all, we know that words we use don't have a 100 percent accurate definition that everyone would agree with, and that there is a lot of ambiguity in language.

In any case, we know what we mean and mostly people understand us. When we sometimes refer to *now* or *the present*, really we mean *the present period*, which could encompass a day, or a year or longer, depending on context and what we wish to convey. The present period in our lives could be the years we have been living in a certain place. *The present decade* or *the present era* are periods in time that are not particularly limited or defined, since they include the past.

We can't get away with it, though—not in a discussion about time. When it comes to time and us, the concept of *now* is a very important one. Essential, even, as we will see. For if we aren't sure about when exactly *now* is and, in fact, whether *the present* is even a period in time, we don't really know where we are and from where we are coming.

The importance of *now* is understood within Eastern thought, that includes New Age in its various strands and links to Buddhism, in which *now* is an intricate part of their basic tenets and is given a central role within people's lives. Here, *now* is all—it is not only when everything takes place, but that the past and future can be accessed only in *the present*. We are always *now*, they say.

Alan Watts, a British lecturer in Eastern philosophy, who spent the latter part of his life extolling the virtues of Buddhism, especially the wonders of *now*, to Western audiences, and who became somewhat of a New Age guru, didn't consider there to be any problems inherent in the notion *the present* when he made this widely quoted statement: "I have realized that the past and future are real illusions, that they exist in the present, which is what there is and all there is."

Watts' opinion is worth examining. Pinning *now* down is the issue. It is, of course, pretty clear that you are reading these words *now* (*the present*). But when did *the present* begin? Surely that beginning is already in the past with every word that your eyes linger on. And we certainly don't know when *the present* will end, because that is still in the future. The faster you read, the faster the words move to the past, while if you read slowly, each word is in your mind longer than *now* that is already gone when you even think about it. It follows, as well, that we don't know how long *the present* takes, since even while we are timing it, part of it is slipping into the past and the rest of it isn't here yet. That happens despite the feeling that an event, say, waiting for a bus, is 'taking a long time.' The thing is that we can really measure the present only when it has gone, when it is the past and no longer *the present*. We can, in other words, know only how long something *took*, not how long it *is taking*.

Yet, despite the fact that it isn't at all clear as to when *the present* (as in 'at the moment') begins or ends, we most certainly know that there is a difference between what we are experiencing and what we experienced in the past. And while we somehow imagine that there is a line, a border, a switchover point, separating the present from the past, we are never actually at that line. Of course we aren't, since there is no such line—time is indivisible: a continuum back to the past running toward the future.

All of those questions and doubts do not mean that we don't have a strong feeling, like Watts, that things are happening at the present time. We function very well under this natural feeling: we know what we mean by *the present* and so does everyone else.

While our feelings are legitimate, can we insist that *now* or *the present* is a real tangible period of time? As we have seen, we can't pinpoint *now*; not only do we not really know when *now* actually occurs, but also that if there is such a moment, it would already be gone, back to the past before we could even say the word.

But wait. Despite all, don't we sometimes capture *now*? When we look at the second hand on a clock or watch, we see it move and it is measuring the movement of the seconds. Every second—every tick—is a *now*, is it not? Actually, it isn't, since when we see it, it is already moving on from the tick that is already in the past. At any point that we say "now!", that point will, in theory, have lasted for a period of time, however short that is (or was, really). The actual *now* is gone long before we can see it, say it or think about it. Marking 'now' on a clock face would be as meaningless as it would be crazy.

Different strokes

It is very human, our willingness to place faith in what appears to be so unquestionably clear as to need no explanation or, indeed, thinking about. It's not surprising that people thought the world was flat; would we not know otherwise, we would still think so. And everything points to there being a present, that things are happening now.

There is no doubt that we feel the strength and reality of our perceptions, that we feel *now* to be a fact. Philosopher Rudolf Carnap recounts Einstein telling him that "the experience of the *now* means something special for man, something essentially different from the past and the future, but this important difference does not and cannot occur within physics."

It's looking at normal, daily occurrences, that makes us reconsider what we so naturally and normally take for granted. When someone talks to us, we see and hear them at that time—'now.' But in light of what we have noted, the scenario is not so clear-cut: what we consider as 'now' takes time. What that means is that although we don't notice it, in fact we begin to see the person before we hear them, owing to the fact that light travels at a faster speed than sound. Putting it another way, we see their lips moving before we hear the sounds they make.

We don't notice this because the distance between us is small; but a sound technician could measure the lapse between sight and sound, even though we are in no need of one to inform us that the flash of lightning gets to us before the sound of thunder, or to confirm that we hear the crunch of a baseball bat hitting the ball after we see it happening. And in the future, the gap between what we hear and what we see will be much more noticeable with the advent of super-human hearing through 'directive beamforming,' the enhancement of acoustic sounds, which will allow us to hear speech more than 30m away. Bearing all that in mind, and if we also take into account that it takes time for transmission of information from the ears and eyes to the brain, it is clear that what we hear and what we see do not reach us together instantly.

When people are looking at an object, say an apple, light takes different lengths of time to get from the apple to each person's eyes according to how far they are from it. Since no person can ever be in exactly the same location as anybody else, each one sees the apple before or after anyone else. Never at the same time, though. This has tremendous repercussions on how we each experience what is around us, since nobody sees things at the same time as anybody else does.

In the normal course of events, that does not matter to those looking at the apple. But it does mean that there is no *now* as such, since each person has their own instantly moving *now*, and that 'what's happening now' is exclusively subjective for each person.

This link between time and location has long been considered by scientists, most famously by Newton and by Einstein (with the latter combining time and space in one dimension, *space-time*). From another perspective, philosophers such as Gottfried Leibniz in the early 18th century (and later, Kant) came to the same conclusion, that time and space are intimately linked. As Leibniz put it, there cannot be two identical objects, since even if they seem similar in all properties, such as shape, color, etc., they would be different because they would be in different locations.

I am taking this a step further. I am demonstrating that no object is identical for everyone, it being different according to who is looking at it and when. In other words, since every object is relative to each person individually, there can be no claim as to which is *the* real one. To put it starkly: since everyone looking at an apple would be seeing it differently, there is no one objective apple; everyone sees it subjectively from their point of view.

There is no getting away from this, let's face it, non-intuitive fact. It doesn't matter how spatially close you are to someone, you can never be in exactly the same position; you will always each see things at a fractionally different time to everyone else—even to yourself. When seeing yourself in a mirror, because it takes time for the light to get to your retina, you don't see yourself as you are, but as you were. You don't even see yourself *now*!

The notion of people doing things together, it would seem, needs a rethink. The point is that there can be no perfect coincidence of any two *presents* (two 'nows'), as no two objects can be present at the same time in the same location. Simply put, it means that we are each in a different *present*!

The fact that nobody is at the same 'now' has concrete practical effects. Take GPS, which relies on people receiving signals from clocks on different satellites. The receiver determines its position on Earth by comparing the different lengths of time the various signals take to reach it. The system depends on the clocks on the satellites all ticking at the same rate. However, they do not. Since time slows down at a high speed, each satellite clock has to have a different perception of 'now.' In order for GPS to be coordinated for us on Earth, engineers have to tinker with the clocks to give each of them their own perception of 'now.'

The whole question of *now* makes us wonder. For a start, it makes us look again at what we consider to be reality. It is popularly claimed that what is real only exists right now, in the present, in contrast to what *was* real in the past and what *may be* real in the future. It is this point—that neither the past nor the future are real—that yoga and meditation guru Isha Kriya uses in order to convince us that we cannot be afraid of either, since they are imaginary. Only *now* is real, he says.

But if each of us has a different *now*, as we have discovered, this surely implies that we each have a different sense of reality from everyone else. Disconcertingly, it kicks the ground from under the whole concept of one reality, especially for *nowists*, who believe that everything happens now and only now. It means that if you point at something and say it is real, it's necessary to realize that it is real in that way for you only; for anyone else, that object is perceived as real differently, being in a different *now*.

This opens up another question that begs to be asked: Which is it that supposedly is happening 'now' in our social world—the voice or the person we see? This is starkly brought into focus when we watch a televised interview with a person far from the interviewer while we wait for the answers to get back. The length of time it takes for the sound of the interviewer's words to get to the interviewee and back again is the cause of those long intervals of silence while the interviewer goes through embarrassingly silent stares at the camera.

When we watch TV, when is what we are watching actually happening? When is it *now*, in other words? Is it when an event is actually taking place, or when we see it on our screens? And what about intentional broadcast delay, so beloved of television producers to help them cover their hindquarters in case of mishaps?

When the broadcasting of live material is intentionally delayed in order to prevent profanity, bloopers, violence or other undesirable material from making it to air, we actually see an event that had taken place in the past. Not only that, but the chances are that we won't see what really happened, since what we actually do see has also been cleansed of mundane problems, such as technical malfunctions or coughing or other inappropriate occurrences. And this is the case not only with mundane problems: when Janet Jackson's right breast was exposed on television at the end of a song with pop star Justin Timberlake during the 2004 Super Bowl broadcast by MTV, it spelt the death knell for live broadcasts, with broadcast delay thereafter becoming the norm.

The present is manipulated in all sorts of ways, some of it for good reasons. Intentional broadcast delay is used in order to take into account different time zones, often to allow a show to air at the same time for the local market in countries with multiple time zones. This so-called time shifting is done quite arbitrarily, so that what the networks consider special events (including award shows) are broadcast live in the rest of the USA while often tape-delayed on the West Coast. For reasons known only to the deciders of such weighty matters, the Academy Awards—the Oscars—are broadcast live in Los Angeles and at the same time in the rest of the USA. Strangely, other events that are actually produced on the West Coast (*American Idol, Dancing With The Stars*) are first broadcast on the East Coast and delayed for the West Coast.

So we can ponder: since people who see a broadcast before we do have a different *now*, do those lucky people who have an earlier *now* experience it in our future or in our past?

And sometimes *now* is relegated to an obscure past, unrelated to what might have been. In President Obama's second inauguration in 2013, it was discovered that the soaring rendition of 'The Star-Spangled Banner' by Beyoncé was mimed (nowadays known as 'lip-synched,' so as to give it a less sullied image). The secret was apparently confirmed by the 'President's Own' Marine Corps Band, which had provided the backing music for the ceremony, and had themselves been accused of miming their performance.

Let's pause for a moment. Following those decidedly non-intuitive views of *now* that result from the relativity of time and location, it should actually be surprising that we are able to communicate with others at all, even in the same room. The fact of there being no *now* as such, and that nobody shares exactly the same *present* as anybody else, justifies our making the seemingly bizarre claim that we are never actually in the same time and place as anybody else!

Despite that, we do, of course, communicate with others, and we can do so because the distances are too small for us to notice that light and sound reach us at different times. Often we do notice, though. When in a crowded place and sounds hit us from all sides, the resultant cacophony—not helped if there is music, as well—is because the sounds are coming from different spots and are assaulting us at different points in time. So while we don't notice time lags from one person talking to us, we find it more difficult to deal with time lags from a number of people.

This is illustrated nicely by our watching a movie in which the sound of the actor's words is not exactly synchronized with their lips. While following the dialogue of one or two actors is slightly disconcerting, it becomes evermore difficult to follow the more characters there are and the longer the conversation among them. We simply cannot deal with the multiple differences in time lag between sight (the speed of light to our eyes from the actors' lips) and sound of the words hitting our ears.

Pinning *now* down

It is, as we can see, hard to know when or what *now* is. In fact, we can't know. How short does *now* have to be before it is already the past? A second? A tenth of a second? The painful truth is that it can never be short enough for, paradoxically, the shorter it is—say, a nanosecond (one thousandth of a microsecond, or 0.000000001 seconds), or Planck time, the smallest unit of time in current theoretical physics (10^{-43} seconds)—the more impossible it is to alight on. *Now*, in other words, is intangible: it is too long to be *now* and also, paradoxically, too short to be *now*. And sometimes the practical action is so short—like snapping one's fingers or a light flashing—that our *now*, when we hear or see it, is after the action has already happened.

The point is that *now* is never now, since it cannot in practice occur at a point that is too short to have duration. *Now* is simply not a length of time.

It goes without saying that when anything happens, it has to take some length of time, however short that may be. That can have an impact. Until fairly recently, financial institutions across London all kept their own slightly different time, and high-frequency trading took advantage of tiny differences in time so that traders received information a fraction of a second before other people, so having an advantage. Traders were able to receive information 1000 times faster than the speed that was accurate within a few milliseconds. The solution to this was to have time measured by an atomic clock and sent on a dedicated line, which settled the issue of accuracy, since all traders receive information within a microsecond, too short to make any practical difference.

To make the concept of *now* even more problematic—as if what we have covered so far isn't enough!—it has been shown that decisions are made in our brain before we act on them. In other words, what we think of as happening 'now' has already happened. This has been demonstrated by neurophysiologist William Grey Walter in the 1960s, showing that there was a fraction of a second delay between the brain making a decision and the person being aware that they were making a decision. Later experimenters seemed to confirm this. In the 1980s, Benjamin Libert of the University of California, San Francisco, showed that there was a surge in brain activity a fraction of a second before his volunteers were aware they were making a decision. Neuroscientist Sam Harris mentions experimenters who found two brain regions that contained information as to what people would do a full 7 to 10 seconds before the decision was consciously made! [2] All of this indicates that what we think is happening *now* is an illusion, and that we are, in fact, 'living in the past.'

As we can see, the problems with the notion of 'now' as a period in time are immense. An attempt at finding the possibility of 'now' by cutting time down to its smallest component unit, is even more disastrous. The principle behind this is that we should be able to understand anything—in this case, time—by examining ever smaller pieces of it, and that when they are reassembled, the small pieces would explain the whole. This cannot work when applied to time, which is a seamless continuum, not consisting of component parts.

If the idea that we cannot get to *now* sounds disturbing, it should do. And it's not new. Already 1600 years ago, St. Augustine let known his worry about not being able to capture the present moment, despite people feeling that they could. Man is "stretched," he said, between the past and the future, so that we are always denied being in the present moment.

The idea that we are never anywhere that is *now* is disconcerting, especially since it leads us to consider the very real possibility there isn't any *now* at all! After all, if we can never time it, what makes us think it exists even in theory? And if, as Arthur Schopenhauer, that most pessimistic of philosophers, would have it that, "Our existence is based solely on the ever-fleeting present," where exactly are we?

And that makes us wonder about the romance that Buddhists and New Agers have with *now*, and its connection to happiness, contentment and wellbeing that it provides. Take, for example, New Age bestselling author and guru Eckhart Tolle, who insists that it is possible to be 'in the now' and, even more amazingly, that *now* is all there is: "The more you focus on time—past and future—the more you miss the Now, the most precious thing there is." In one fell swoop, Tolle tells us that we shouldn't even think about time since by doing so we miss the *now*. In other words, we are in the *now* only when we don't think about it as part of time.

Tolle's statement is interesting, if you don't think about it. "Why is Now the most precious thing?" he asks us. To which he answers, "Firstly, because it is the only thing. It's all there is... Life is now. There was never a time when your life was not now, nor will there ever be."[3] In light of what we have seen about the impossibility of even a fleeting appearance of *now*, the fact that statements like Tolle's resonate with many people, must make us wonder why. If the *now* being referred to is supposedly situated in time, then, as we have seen, it is extremely problematic to imagine where it is or what it is, and to accept it as existing within time at all. For if *now* is deemed as the only time that exists, it would, in fact, not be time at all; it would be, as Tolle intimates, without length—a durationless time that is always present. How, though, is it always present if it doesn't last any length of time? We are left with an entity that cannot be measured and that is not part of our world.

Yet, problems of time do not seem to affect Tolle too much, since he is selling *now* as a gateway to happiness while being an everlasting part of time. It is the search for instant and eternal satisfaction. Pepsi obviously agrees; their slogan 'Live for the now' certainly catches the spirit of the times. And while it is perhaps surprising that the most visited place in the world is a shopping mall—the 6-million square-foot Dubai Mall—it makes sense that it is marketed as 'the center of now.'

This is an important point. If *now* were touted as a spiritual entity only, it could be acceptable on those terms. But that is not it. The justification for *now* leans on a supposedly rational link to time, its spiritual *raison d'être* based on a period in time. As for how and why 'living in the now' is deemed to be connected to our wellbeing, let's leave that to later, when we look at New Age, self-help, and attendant hopes and aspirations.

In the meantime, before we roam through different aspects of time so important to us, that include memory, free will, self, déjà vu, art, hope and much else, let's talk about time itself.

Is time real?

A point often made to me by people who heard I was writing about time was that time is a notion invented by humans, that time doesn't really exist outside of our clocks and timing devices. At first blush, that doesn't sound unreasonable; after all, time is not something that we can actually see, and nor are we really aware of time unless we stop to think about it.

The question of what time is, and whether it is something real or illusory, is an ancient one. From the Greeks to Newton to Einstein to the present day, time has been discussed and argued about. It is so much part of our lives and everything we do—Thomas Mann maintains that the perception of time is so closely bound up with the consciousness of life "that one may not be weakened without the other suffering a sensible impairment"[4]—that it is not unreasonable for time to be the subject of so much discussion.

Yet, despite being so ubiquitous, somewhat like air, always around, time is easily dismissed as not real as such, being merely a function of 'something that clocks measure for us.' Unlike air, which we know to be essential, time somehow does not strike people as having the same vital importance. And unlike food, necessary for our physical existence, time goes on whether humans partake of it or not. So while lack of food would be noticeable, painful and eventually fatal, lack of time, though sometimes irritably apparent, is not thought of as having the same dire consequences.

The point that time is unreal, that it does not exist as such and is an illusion, is made most consistently by Buddhists and New Agers, as for example, Alan Watts, who states that "time is a social institution and not a physical reality." There is no such thing as time in the natural world, he says, but rather, there is 'rhythm.'[5]

There is, though, no doubt that we feel the passing of time, as do animals who function within it and cannot be accused of thinking of time as something measured by clocks. Tracker dogs can 'smell time' when they come across footprints and know which direction the person has gone. They do so by the difference in odor concentration, which is weaker at the earlier footsteps and stronger at the most recent ones. So they can tell what was there more recently and also what is about to arrive from smells in the wind.[6]

We don't need a clock to determine when we eat; we are simply hungry, sufficient time having passed from when we ate last for us to feel the need to eat again. A look at an old photograph of ourselves or of anyone we know, will show, sometimes quite shockingly, that time has passed. The sun's movement across the sky from morning through afternoon to evening will show us that time is passing and has passed. If and when it occurs, a midlife crisis indicates that time has passed, the physiological and mental changes as stark reminders. And what is a 'biological clock' if not the passing of time?

Yet it is significant that Alan Watts refers to 'rhythm' rather than 'time,' since his big issue—the question of *now*—is a spiritual one, not a physical or scientific one. *Now* as timeless is the point. Watts puts it like this: "If you listen and pay close attention to what is, you will discover that there is no past, no future, and no one listening. You cannot hear yourself listening. You live in the eternal now and you are that. It is really extremely simple, and that is the way it is."

It is simple, though, only if *now* is not considered as part of time. And even simpler without the existence of time itself, we are told. But without time, the notion of eternity (and within it a so-called 'eternal *now*') is extremely suspect. And dismissing the past and the future so readily flies against changes occurring within nature.

The American theoretical scientist Brian Greene defines time as "the mechanism by which you can notice change." Time is change, which is part of nature itself. Time is woven into the fabric of nature. So while it is true that the way we refer to units of time and how we measure them—'telling the time'—is convention within each language that contains them, time as change does not need to be measured to make it real. Time and space are as real as anything that we can see and touch.

Civilization would have quickly died out if the first humans did not link the passing of the sun with the food that grew naturally, so linking changing of the seasons with the sun and moon. The concept of time and change with the ability to create ways of measuring them goes back some 10,000 years. Evidence of a giant 'year clock' from 8000 BC in Banchory in Scotland, is a set of pits aligned with key astronomical events that was capable of tracking the passing of lunar months and linking these to the changing of the seasons, so enabling the middle Stone Age hunter-gatherers to prepare for changes in the food supply. Most amazing was the ability to track the midwinter sun, thus predating by some 5000 years the first formal calendars created in Mesopotamia.

It was not time that was 'created' by these ancient people, but that they found a way of measuring events of nature, and predicting when they would occur and reoccur. They quantified time, in other words. What they invented was the ability to 'tell time.'

While harnessing time through natural elements such as the moon and the sun, people must have been early on aware of what we now call 'the internal clock.' Like plants in their way, we wake up and go to sleep according to an inner mechanism that is influenced by the sun. But even without the sun showing the way, the body contains clock genes, which, says Nicolas

Cermakian, of McGill University, Montreal, "are what drive our biological clocks, and these genes are active in all of our organs." Animal studies have shown that our central clock in the brain sends signals to the clocks in our other organs. And that clock can be moved by a certain type of steroid as a treatment for jet lag.

Our internal biological clock doesn't always work properly. Just like mechanical clocks, ours sometimes breaks down or goes haywire, which happens when we suffer from jetlag and our body works in a different 24-hour time span to where the body is actually located. Now while we generally catch up within a few days, there are times when the body is more seriously affected. Malfunctioning of one's biological clock, called 'sleep and circadian rhythm disruption,' or SCRD, can lead to a litany of mental illnesses. Conditions such as schizophrenia, bipolar disorder and depression are part of it, while it has been known for years that shift workers tend to have higher rates of Type 2 diabetes, heart disease, infection, cancer and mental illness, and are more prone to accidents.

Time, then, exists not merely because people think it does. Time happens. Things happen. Everything changes. The way we think of time and relate to it may be subjective ('a watched kettle never boils' or, as the British writer Claudia Hammond puts it: "A watched pot never seems to boil, but go and check your emails and it will be boiling over before you know it."), but time itself is an essential part of what there is. Time is as much qualitative as it is quantitative: we should not confuse measurement of time—definitely a human invention—with time itself, an intrinsic part of human nature. The fact is that everyone experiences cause and effect in that order, rather than effect before the cause. We were born after our parents and not before them. People die after they are born, not the other way round. Not only do all humans experience things happening—change—but that cause and effect is universal. The 'arrow of time,' as the astrophysicist

Arthur Eddington first called it in 1927, is progressive change from the past through our perception of the present to the future for us all, whether we think about it or not.

Time is movement and things happening. If nothing happened, there would be no time. If time stopped, would we stop aging? And there is no such thing as nothing happening. Things happen just by humans being around. Even if we did nothing (whatever that could possibly mean) and we were in a completely soundless, lightless room, things would be happening by the very virtue of our being alive: heart, lungs, stomach, pulse, kidneys, liver, etc., all are functioning. And we would hear things in our ears: ourselves actually living.

To divorce time from the world is an impossibility. Where there is space, there is time (the space-time of Einstein's two theories of relativity); anything that has location, in other words, must be part of time. Since time is an integral part of everything, it connects time to 'existence.' There are, though, dissenting voices. Philosopher Roman Altshuler says that a rubber duck, for example, exists intrinsically without any connection to time. He does not explain how, where and in what way any object could exist outside of time. "Nothing in the metaphysical structure of a rubber ducky requires that it move in one way rather than another, and if the universe were suddenly frozen at a single moment in time, rubber duckies would still exist." The point, though, is not how the duck moves, but what a changeless duck could possibly be, and how a changeless world could exist at all. A changeless world is, in fact, an oxymoron, as is a changeless anything.

Where there is something, there is time. To put it simply, time is stuff. Without time, there is no stuff, and vice versa. Only Nothing is timeless; everything else is time-full, and there is no doubt, as Emile Durkheim said, "that we cannot think of objects which are not in time and space."[7] This is different to what Immanuel Kant thought. His belief in the supremacy of time, that there can never be an absence of time, was so strong that as far as he was concerned, time would be present *even if there were no objects*. That is a problematic position, since we would need to ask: On what would time work? Where would there be change—what would change—if there was nothing around? So if we ask whether we could live without time, the answer is that not only would there be no life, there would be nothing without time. There would be no us without time.

How would the New Age *now* as a purely mystical element, not connected to time at all, work? A mystical *now* is not really a moment as such, so that when we are encouraged to "realize the precious nature of each moment," as does the New Age author Rick Singer,[8] it is because of the supposed transient nature of the moment. Seize the moment—*carpe diem*—is his advice. But if *now* is eternal and durationless ("This moment is all we are ever guaranteed. Life equals this moment and nothing more,"), this doesn't square with the idea that the moment will move on if we don't treasure it.

The notion of 'moving on' is itself interesting and fairly recent as a general solution to pain, angst, mourning and a host of events that we should 'put behind us.' We are constantly exhorted to move on, to accept the situation and to achieve closure, as if dwelling on the past is somehow unhealthy. Yet, 'moving on' is a contradiction to the notion of *now*, both as a part of time and as a mystical element. In his bestselling book *Wherever You Go, There You Are*, Jon Kabat-Zinn exhorts us to "actually *feel* the present moment, to see it in its fullness, to hold it in awareness, and thereby come to know and understand it better. Only then can we accept the truth of this moment in our life, learn from it, and move on." But from where are we supposed to move on? Elsewhere he says that "The only time you ever have in which to learn anything or see anything or feel anything, or express any feeling or emotion, or respond to an event, or grow, or heal, is this moment, because this is the only moment any of us ever gets. You're only here now; you're only alive in this moment." If *now* ('the moment') is all that there is, how can we move on from it? To where, if the future purportedly does not exist, and from where, if the past does not exist?

Not only that, but the very notion of 'seizing the moment' belies the concept of time as a continuum that cannot be split into elements that can be 'seized.'

With all the questions and confusion with whether *now* is part of time or whether it is a purely spiritual element, it is not surprising that *nowists* are very ambivalent about the existence of time itself.

Where misapprehensions occur as well is when time and its measurement are confused. An example is that of the physicist Julian Barbour, a major proponent of the idea that time does not exist, who states that "even Einstein failed to challenge the concept of time as a measure of change." But that is the point. Time as a measure of change is a human conception. Time is change itself, which occurs whether we measure it or not.

The notion that time doesn't exist except as 'now,' as claimed by Buddhists and New Agers, is a strange one considering how concerned we are with time, how much of our lives are taken up by it and how dependent we are on getting things done 'on time.' We can perceive time through the changes that occur, and feel it 'slipping through our fingers.'

In *The Hitchhiker's Guide to the Galaxy*, Ford Prefect says that "Time is an illusion. Lunch time doubly so." Perhaps. In our world, a double martini at lunch could well give us the feeling that time (and much else) is an illusion and does not really exist. On the other hand, we could imagine time to be "nature's way of stopping everything happening all at once," as quoted by the theoretical physicist John Wheeler.[9]

Always

Now is a time killer. That is the problem with the durationless *now*, as Aristotle noted almost two thousand years ago: "As, if the now had remained the same, time would not have existed." An unchanging *now* within a changing time is an impossibility; they cannot exist in tandem.

Yet, no matter whether a durationless *now* is possible or not, it does have the attraction of eternity that it conjures up. After all, if *now* is all there is, then why worry about a nonexistent future?

Eternity is always a good sell. It's not for nothing that the promise of eternity in some form or other is the common denominator of all religions, faiths, beliefs and cults. Usually, it refers to after death. Often, though, it is about an eternity that transcends time, that which the influential medieval Church father, Augustine of Hippo, 1600 years ago referred to as God's eternal *now* moment, in which there was no past, present or future. So when Eckhart Tolle says, "Eternity is the living reality of who you are," which is his description of now, he hits a familiar and beguiling nerve, and definitely a more attractive life proposition than the timeless afterlife described by Kyriacos C. Markides, a writer on Christian mysticism: "Also, when we speak of eternity, we are not speaking of a state to which we can apply our categories of time. Hence we are not in a position to say that eternity can last even for a single second."

How much more comfortingly attractive is the mythologist Joseph Campbell's exhortation not to worry about what happens after death, since everything happens in the here and now: "The experience of eternity right here and now is the function of life. Heaven is not the place to have the experience; here is the place to have the experience."

As manifestation of a widespread desire for eternal adolescence, sex, drugs and wild abandonment are accepted as ways toward the goal of eternity, as were the antics of Michael

Jackson, who reshaped himself physically to remain trapped with his high speaking voice within his private Neverland Ranch, California. Increasing numbers of people on the verge of old age dress as they did when they were young, reluctant to acknowledge that their youth has passed.

It is no wonder that celebrities and those obsessed with the urge for eternal youth go for the lure of *now*. "There is nothing inevitable about aging," we are told by Deepak Chopra, a doctor at Tufts and Boston University before turning to meditation and then bringing out, in 1993, his breakthrough book, *Ageless Body, Timeless Mind: The Quantum Alternative to Growing Old.* In one soothing voice, East met West, the mind met the body, and the aging boomers met their age-defying guru. It is not surprising that the notion of agelessness hits the right buttons. Concentrating on 'now' has been given legitimacy within a spiritual dimension and a metaphysical grounding.

The notion that time can be slowed, that eternity—at least as far as eternal youth is concerned—can somehow be attained, is exemplified in the rapid expansion of the 'anti-aging' industry in the West, that peddles an airbrushed vision of a world in which aging or mortality can be almost indefinitely deferred by the dutiful and desperate ingestion of supplements and restless application of pseudo-scientific skin treatments.

Eternal youth is sought after by the super-rich, who seem to go through every sort of pain and discomfort to achieve it. The paralyzing discomfort of being wrapped in cellophane and being squeezed into a tube for 'detox' and 'oxygenation' is small fish compared to the four times per year for blood tests and six-monthly 'blood work,' not to mention the regular Botox and injections of vitamins and occasional intravenous drips. This is only part of what Tom Bower details in his biography of the TV personality Simon Cowell. What it cannot offer, of course, is any guaranteed change to the final outcome.

And yet eternity is a sought-after commodity, brought into

mainstream consciousness and exemplified most dramatically by pop groups popular in the 1960s, who are still going strong, with entertainers in their 70s performing the same numbers they did 50 years earlier. Most dramatically, it is the Rolling Stones that hit us in the face with their antics that seem to defy old age. Despite the obvious physical ravages of age, the Stones give the illusion of perpetual motion within the disjunction of present, past, and memory of what still seems to be. When they first performed 'Time is on My Side' in 1964, being in their twenties, they were right. Presumably, they didn't mean it to be that they would still be performing it 50 years later. Ironically, they also still perform 'It's All Over Now'. It doesn't seem to be, though. Not yet.

Unlike Bob Dylan and Leonard Cohen, both holdovers from the 1960s, who continued to write and perform songs that reflected the people they had become, as much as the people they once were, the Rolling Stones' music is essentially and inescapably defined by the times in which it was made, so perpetuating the myth that time could possibly be tamed, perhaps even frozen. And yet... If time is on our side, looking at them would be a reminder of William Wordsworth's "bliss was it in that dawn to be alive, but to be young was very heaven!"

Language helps, as well. Baby boomers—those born during and just after the war (WWII is always called 'the war')—are no longer associated with 'old age,' but in repackaged and evermore soothing and fast-changing euphemisms, such as 'the golden years' and 'the third age.' And nor are time spans what they used to be: if 'life begins at forty' was reasonable for earlier generations, this is now more realistically sixty, an age when people used to be close to retirement but nowadays find a new spurt of life to begin a new career or become newly creative. So when the late David Bowie brought out a new single at the beginning of 2013 at the age of 68 after a 10-year break, it was par for the course.

With all that, it is no wonder that people have a new belief in the possibility of things going on and on, and that what we are like today will always be. Witnessing the antics of ageless pop stars and Botoxed celebrities enables a fantasy of endless youth and health, in which old age can seem to be an inexplicably bad lifestyle choice, like eating junk food or smoking, that you can avoid if you're well educated or cool enough.

It's not as if we aren't aware of disease or don't come into contact with it among people who are close to us; it's that it always happens to someone else. This is especially so if we each believe we are 'living in the now,' allowing us to float past the bad stuff that does not affect us. We can't get enough of explosive gory death onscreen, but we look very differently at scenes of the slow death that is of the creeping, gray, incontinent kind. A combination of 'seizing the moment,' 'mindfulness' and 'self-empowerment' will, we are told, keep the illusion of time at bay.

We live in an age of Peter Pan. Or at least more and more people believe they do. Baby boomers don't think they are getting old, and the ones who can afford it refuse to think so. It is not eternal childhood that they seek, however, but eternal adolescence. Neverland has become Alwaysland. Encouraged by New Age exhortations, such as those in the interestingly titled book *Making Time: Why Time Seems to Pass at Different Speeds and How to Control it*:[10] "It is actually possible to *expand* time, to alter our perceptions to experience *more* of it."

It's an escape, of course, from facing the inevitability of time's forward march. It was his last refuge from reality when Winston, in George Orwell's *1984*, said that "Nothing exists but an endless present." As a way of surviving suppression and oppression, it helped—but very temporarily.

The notion of durationless time—not time at all, in fact—is definitely attractive. It makes the eternity of afterlife quite bearable and *now* a living eternity, an absolute joy. Perhaps Yossarian, in *Catch 22*, summed it up best when he said that he "had decided to live forever or die in the attempt."

It must make us wonder, though. If it is a good idea for post-death, would eternity be so great if it replaced death? How wonderful would it be living the eternal *now*?

In his book *Timequake,* Kurt Vonnegut mentions a scene from George Bernard Shaw's play *Back to Methuselah*: Adam and Eve are waiting for God's annual visit to hear that everything is okay and that they are grateful. This had been going on for hundreds of years, but this time they wanted to tell him something new. So when he turns up, he again expected to hear that everything is great, since obviously all that he had created was perfect. Adam and Eve are somewhat nervous, but needed to say what was on their minds. They tell God that everything is absolutely wonderful, but that it would be even better if they could know that it was going to *end* sometime.

Perhaps that was why they ate the apple eventually. As author Chuck Palahniuk asks: "Did perpetual happiness in the Garden of Eden maybe get so boring that eating the apple was justified?"

An old cartoon by Gahan Wilson captures the idea. Two middle-aged angels, newly arrived in heaven, are playing harps while perched on a cloud. One, his face a mask of dismay, asks the other, "You mean this goes on forever?"

Vonnegut's *Timequake* gives a different slant to the prospect of eternity. It is a situation where a 'timequake' in 2001 causes everyone in the world to go 10 years back in time, to 1991. From that point, the world begins an exact copy of the events that had already happened. Memory of the past was not deleted, though. What is interesting in this scenario is that because everyone is able to remember everything that was going to happen during the next 10 years, they had no free will at all. With memory becoming the present and future, people didn't need to do anything in order for things to happen. Life, in fact, became effortless, even boring in its predictability within the inevitable tragic consequences of a miserable, unavoidable, deterministic cycle.

The notion of a living eternity is at least as fascinating as an eternal afterlife. It certainly has more variations, as can be seen in the imaginative treatment afforded it through novels, movies and theatre. Philosophers have, of course, gone into it in depth, but very few as depressingly as Friedrich Nietzsche in the late 19th century. That he should look at it in that way is not surprising, since his life was mired in depression and increasing bad health, all the while espousing his philosophy that reflected a desire to challenge the basis of philosophical thought of the period.

Imagine, Nietzsche asks of us, if everything that happens in the universe, including our lives down to the slightest detail, will eventually repeat itself, in a cycle that plays out for all eternity. "This life as you now live it and have lived it you will have to live again and innumerable times again; and there will be nothing new in it, but every pain and every joy and every thought and sigh and everything unspeakably small or great in your life must return to you, all in the same succession and sequence..."[11] Not something to look forward to, and certainly not better as an alternative to the linear birth-to-death events.

Eternity lite

There are all sorts of eternities, though. For the exercise, let's consider the pleasure of having our cake and eating it: repeating only those events and circumstances that are pleasant, that we would want to have again and again.

It sounds like a good idea. But is it? What would wonderful occasions be like if you knew they would be repeated? Perhaps not as good as we think. For a start, the events themselves wouldn't be quite as poignant at the time if we knew they would happen again. Surely a large part of the feeling we all have, even if unconscious, is that whatever we do, we will not do it again—or at least not in the same way; we may not often think about it, but our experiences are colored by the fact that each is unique.

Unique not just because it is different from anything that will come afterward, but because of our awareness that life itself is time limited.

Other problems pop up at the thought of choosing to repeat only our pleasant experiences. Just think of the decisions one would have to make as to which experiences we choose and which ones we decide to let slip away. Filled with regrets we'd be about the events we didn't choose, while being left with repeats of events that would inevitably turn out to be not so great the second or fifth time round. Perhaps that is why 'never go back' to what was so marvelous is such wise advice; a second visit to that ideal holiday place is never quite as good as it was the first time, since time allows nothing to stay as it is or, in fact, as it was.

Nevertheless, many of us would have fantasized about a perfect situation continuing forever. It's a common wish made without much thought. "It's the thing we've all been waiting for and praying for, for so many years. It's been a surreal experience and you just don't want it to end," said Will Murray, father of Andy Murray, the first British Wimbledon men's singles tennis champion in 77 years. And who hasn't wished that that sunset with a loved one would just go on and on—that savor-the-moment contentment? As the American educator Henry Van Dyke pointed out so aptly: "For those who love, time is eternity."

The urge to stop time is real and it is all-consuming, as in these heartfelt words by the 19th-century French poet Alphonse de Lamartine: "O time, suspend your flow, these perfect hours, and stay!" A legend about Venice says that lovers will be granted eternal love if they kiss at sunset under the Bridge of Sighs while the bells of the Campanile toll.

But just imagine it. How long before the excitement would wear off? Not only of everlasting romantic love, but of more mundane scenarios. What about the ideal of not needing to work? It sounds like paradise on Earth. For some, the attraction would be replacing the grind of daily work with lying on a beach on an exotic island, drinking cocktails and reading when in the mood. But what then? Sooner or later, this will become boring. Having no aim is frustrating. It eventually becomes impossible.

The idea, of course, is for it not to wear off, for the novelty to be permanent. But it can't be. A novelty has to be fresh; otherwise it isn't novel. Even the idea of working in the paradise of an exotic location, away from an office, loses its luster in time. Many Westerners have tried it as more and more jobs are becoming portable and possible to do at a digital distance—not just Web designers and freelance writers, but fashion designers, photographers, models, marketers and even remote-working medical doctors have settled down in places that have been designed for it, such as Ubud in Bali.

What can be more idyllic than 'location independence,' also known as 'digital nomadism'? For some, apparently, a few weeks of that theoretical paradise palls. As journalist Anna Hart reported, "Six weeks of that lifestyle was enough. The freedom to work in exotic surroundings is thrilling, but it quickly becomes isolating, uninspiring, lonely and—depending on the Internet speed—frustrating."

Anything new becomes old in time. When it becomes a continuation of what you remember, it's no longer a novelty. As Leonard Cohen in *Beautiful Losers* put it: "How can I begin anything new with all of yesterday in me?"

Yet, we tend to think in long-term phases, as if things won't change. The tattoos that people inscribe on themselves are outward signs of such wishful thinking. Tattoos, that used to be the preserve of sailors, prisoners and punks, have made it big time to anyone who is anyone (and even those who aren't). Robbie Williams, Johnny Depp, David Beckham and Mike Tyson, not to mention Kate Moss and Angelina Jolie, are walking coloring books and role models. One out of 10 Britons has a tattoo, as do 45 million Americans (40 percent of 26–to 40-year-olds and

36 percent of those aged 18–25). You have to be pretty sure that your feelings won't change to have a lover's name tattooed on your arm. Not to mention that the ideas of youth do not wear well in later age, as exemplified by that-seemed-a-good-idea-at-the-time crocodile tattoo round the nipple of a now-60-year-old man sunning himself on a beach. It didn't seem to him that *now* would ever become *then*.

Strangely, although we know that things will change, we act as if it won't happen. It is as if some sort of denial process is in place that gives us the delusion of *now* as a permanently-suspended-in-place movement of time.

If we were offered an everlasting life with good youthful health, would we take it? And let's say that the offer contained a proviso that if we did take it, we could never change our minds later and reverse our choice. How many people would go for it? Strangely, most people would decide not to take up the offer. Ignoring those who altruistically claim that it would lead to overpopulation, it seems that the thought of living forever is overbearing. Somehow, humans need to know that life is finite, that everything comes to an end. The knowledge of one's earthly mortality is engrained, it seems. So much so that the alternative is not an option, even if offered. How strange are we humans.

"Eternity can be a curse," the dying old man speculates in the 2013 movie *The Wolverine*. "A man can run out of things to live for."

Is time an illusion?

That time is an illusion is put out often. But if that is so, if time, ubiquitous as change within nature, is an illusion, then all of nature would be an illusion.

We could, of course, entertain the fact that the world around us is an illusion, as posited by Hinduism, or a falsehood thrust into our heads by mad scientists or evil demons. Perhaps the words in front of you are an illusion. But even if that were so, we cannot cast away time so easily. Time passes whether what there is, is an illusion or not. Where there is thought and awareness, there is awareness of time passing. As a follow-up to Descartes' "I think, therefore I am," we could rightfully claim that "where I am, there is time." Nothing can exist without time, not even illusions.

If we stop for a moment to think about it, is it not a strange idea that everything around us, including time, should be considered an illusion, while *now* is touted as not being an illusion? Yet *nowists* suggest that we ignore our own intuitions about the passing of time, while accepting that one intuition—that of *now*—is real!

And so it is with the connection between time and space, that are like two sides of the same coin (even if the coin is an illusion). Why would time be more an illusion than, say, distance? Going from point A to point B is a physical act, yet intimately connected to time. Getting to a party in two days' time will require us to move a physical distance. Once we are there, it will be our 'here' and 'now,' instead of 'there' and 'then.'

Time passes; that is its nature. Yet, we don't actually see or feel time passing; we notice change only after it has taken place. It's the *result* of time that we are witness to. We don't see seeds sprouting or children growing; we see that seeds have sprouted and that children have grown. If we wanted to, we could, of course, see those happening on film or video that had been sped up. However, we would not be seeing time actually passing, but rather a recording of the event (of time having passed): we would still be after the event having actually taken place, not while it was happening. And just as relevant: whether we feel time actually passing, or watch it sped up at a later date, each one of us is already older, not having noticed our own aging, either. Time in its relentless march forward makes sure that we are always older than we've ever been before and that it's the youngest we are ever going to get. The delusion of *now* saves us from the uncomfortable consequences of that thought.

We are regularly reminded of time having passed, often shocked at seeing someone after a lengthy absence and noting how much that person has changed. We would not have noticed time passing and changes taking place had we been seeing that person constantly.

We come face to face with this when we look at a photo of ourselves. Whenever the photo was taken, the *me* looking at it is not the same *me* as in the photo, if only for the fact that time has passed. The *me* who is looking at the photo realizes that time has passed by noticing how different the younger *me* is. As the comedian Mitch Hedberg said incisively, "One time a guy handed me a picture and said 'here's a picture of me when I was younger.' Every picture is of you when you were younger."

And so it is with time and space: *now* and location. Because it instantly and constantly moves to the past, *now* is always changing, so that informing someone that 'I am here now,' is actually a perpetually changing description of where you are. It is, in fact, a description of where you were.

> "Isn't it funny how day by day nothing changes, but when you look back everything is different."
> C. S. Lewis

Despite all, the notion that time is an illusion persists, with science sometimes cited to back this up. Einstein, so we are told, said as much. He didn't, though. His actual words, in a consoling letter written to the family on the occasion of his friend Michele Besso's death, were: "People like us, who believe in physics, know that the distinction between past, present, and future is only a stubbornly persistent illusion." It is clear that Einstein was referring to the illusion that there is a distinction between past, present and future—not that time was an illusion, but just that the division within time was.

This common misconception—that dismissing the veracity of the past, present, future division entails a dismissal of time itself—is often perpetuated through a genuine misunderstanding. Sometimes, though, one cannot help suspecting that wrong information is imparted as a means of pushing a specific agenda. It certainly helps that viral repetition of information across the Internet has the effect of giving a stamp of legitimacy to these ideas.

In the bestselling self-help book (and video) *The Secret*, which claims that we can do anything we set our minds to, the authors back up their claims through physicists who have ostensibly proven that time does not exist: "Time is just an illusion. Einstein told us that."[12] No, he didn't, as we have seen. Not only didn't Einstein state that time is an illusion, as we can see from the actual quote earlier, but, in fact, his work was based on aspects of time and space (space-time) as an integral part of the universe. What were his theories of relativity if not on how time works? As the British physicist Julian Barbour—who posits the notion that time does not exist!—puts it: "Einstein, in a way, makes time something real—with the idea of space-time, he makes it as real as space."[13]

And while we are at it, it's worth looking at another claim in *The Secret*, based on what Einstein purportedly said: "What quantum physicists and Einstein tell us is that everything is happening simultaneously." No, they don't. The result of Einstein's work is that nothing can travel faster than the speed of light, that time is not an absolute quantity and that every point in space-time is limited in the way it can communicate with other regions of space-time. So the idea that everything happens simultaneously is exactly counter to the results that Einstein developed (and, incidentally, supports our claim that no two events happen simultaneously). The notion that the mind produces energy that connects it to other minds cannot be gleaned from Einstein's theories.

And that is what we have already gleaned from an examination as to whether or not *now* could possibly be real in any way. Obviously, if the present, the past and the future are considered to be illusions, *now* would also be an illusion. As already pointed out, it takes time for the brain to process all that enters the eyes and the ears, so the strange fact is that we are actually living around as much as half a second in the past. The *now* that we are experiencing, we actually aren't. Mostly it doesn't matter, but often we need to make allowances for that lost half-second, as, for example, when we go to catch a ball, our brain will predict where the ball will be so that we can catch it.

The brain is constantly making allowances for this delay in visual images and sound. Musicians playing to an audience get feedback from the equipment facing them on the stage, so that what they are playing can be bounced back at them as instantaneously as possible. That it isn't actually instantaneous is owing to the delay in the sound getting back to them. They hear what they are playing and/or singing, but do so a second or so after they have released the sound. In addition to the mechanical feedback is the human feedback from the audience, which has a longer delay depending on how large the audience and venue are.

So while the divisions—including *now*—within time are an illusion, time itself isn't, as can be seen by how we apply it in our lives and consciously are able to use it, aware as we are of the inherent disharmony between our perception and the physical 'reality' of time. It has been shown, for example, that humans have a remarkable ability to keep track of time. Luke Jones of the University of Manchester, among others, has shown that we're uncannily good at telling the difference between durations of sound, down to one-tenths of a second. Strangely, though, we are not at all good at predicting how long it's going to take us to do something, even if it's something we've done before. (Apparently, we always underestimate.) He says that "on a functional level, the brain is acting like a clock. We have some sort of internal clock for measuring events." People can count in seconds pretty accurately, whether it be through chanting 'one Mississippi, two Mississippi' or whatever trick they use, but they can't do this if they are given another task to do at the same time; for example, add up a row of numbers.[14] Interestingly, it was also found in another study that depressed people are far more accurate in time estimating than their happier peers, possibly because mildly depressed people focus their attention on time and less on external influences, and therefore have clarity of thought—a phenomenon known as 'depressive realism.'[15]

There is, as already pointed out, a difference between timing something (measuring time) and what time itself is (the concept of time). Confusion between them persists, though; for example, in sensational stories about hitherto isolated tribes, who, we are told, have no concept of time. So when the respected journal *Language and Cognition* reported an Amazonian tribe, the Amondawa, where words like 'month' and 'year' don't exist, and that age is marked by a change of name, while very interesting, this does not necessarily mean that the tribe has no concept of time, as claimed in the article. Whereas it is tempting to think of time that is naturally measured in units (seconds, minutes, hours, weeks, months...), it so happens that much of the world does not measure time in the same way, but rather according to its connection with natural phenomena.

Despite having no words for such concepts, the Amondawa people recognize and take advantage of divisions of day and night, and rainy and dry seasons. Does this not indicate that they do have a concept of time passing? They simply don't group the different days together into weeks, months or years and, hence, don't need names for the different days. The fact that they know that there is a new day after the night, and a new one after the next night, we can assume that they know usefully enough how long they have before the dark catches them. They will know that sunset is preceded with predawn sounds from animals by the way birds and other animals act, informing them of when to do certain tasks.

It goes without saying that language reflects the way people relate to time. It is not surprising that the Sioux, for whom 'being on time' has no meaning as such, have no word for 'late' or 'waiting.' Now while the Spanish, albeit Europeans and not Native Americans, do have a word for 'late,' they are popularly known to be rather cavalier with the whole concept. As such, they may jocularly use the word '*mañana*,' which means 'tomorrow' and also 'morning,' in its lackadaisical, whimsical sense, to infer that tomorrow may never come. The Arabic '*bokra*,' which has the same meaning ('tomorrow' and 'morning'), is used in exactly the same way. Time can be fluid indeed.

To those who suggest that time is not real or is an illusion, here's a question: In which way is time less real than, say, a tree? To the answer that we can touch a tree (a physical object), but we cannot touch time (an abstract concept), it should be mentioned that we cannot touch thoughts or dreams or gravity either, even though we know where and when they are physically reflected in the brain. Surely we wouldn't say that they aren't real, would we? And we come into contact with time, that affects us in every way, as often as we do with gravity, and a lot more often than trees or even dreams.

Much of the problem and discussion about time is that we don't really know what it is and find it difficult to define. Now if we were to say, "We do not understand this thing called 'time,'" that would be a concrete statement. But from not understanding time to stating that it doesn't exist, or that it is not real, or that it is an illusion, is unwarranted. The world is full of things that we don't understand but we still accept them as existing, in so far as we can talk about them. Consciousness, mind, personality, gravity, light, space are just a few examples that fall into that category.

And, yes, time is relative. And so, in fact, is everything. There isn't anything that isn't relative to everything else: you sitting down reading this book, the bed in your room, the place in which you live: all are moving because the Earth is moving. Here is how it is sung by Eric Idle in the Monty Python movie *The Meaning of Life*:

Just remember that you're standing on a planet that's evolving
And revolving at 900 miles an hour
That's orbiting at 19 miles a second, so it's reckoned
A sun that is the source of all our power
The sun and you and me, and all the stars that we can see
Are moving at a million miles a day
In an outer spiral arm, at 40,000 miles an hour
Of the galaxy we call the Milky Way

Since 1981, when the lyrics were written, although expanding scientific research and observation have shown the original figures to be slightly off, the principle that everything is moving relative to everything else is clear.

With everything being relative and always in motion, we can't pinpoint anything, since it is only roughly where we say it is. Obviously, that applies as much to location as it does to time. So while we feel that what we see and what we do are in *the present,* it is equally clear that our perception is always slightly off. Just as our stones hit the wall at approximately the same time, so the words you are reading are only approximately where you think they are in space and time. It is never exactly so.

Old man river

Whether time is not an illusion—and even if we think it is!—it affects everything around us. The way we live our lives is the way we relate to time and behave within it. In fact, everything we do is with time as a backdrop, never mind how often we do or don't think about it.

Before we even start on that, and no matter how we imagine time works, we know one thing for certain: that when our discussion ends, it will do so after we have begun it, not the other way round. That is another way of pointing out something obvious: time always moves forward. What is so figuratively, yet aptly, known as *the arrow of time* moves forward, not back. We remember the past and not the future. Animals are born before they die, a plate is broken after it was whole, omelettes are prepared after the eggs are cracked open—and all of these are irreversible.

The metaphor of an arrow is eminently appropriate to show the irreversible linear direction of time. Past, present, future. It is what we expect: the experience of 'next,' and the inevitability of the next after that. Whether we look back to the past or forward to the future, we move—or time moves us—from birth to death.

Theories in science, though, consider the direction of time more liberally. While acknowledging that time moves linearly from the past to the future, physicists maintain that there is no scientific reason for time not to move the other way: 'backward' from the future to the past. Since in practical terms it doesn't happen, such a theoretical possibility leaves many big question marks. The fact is that we have come across it so far only in fiction. In J. K. Rowling's *Harry Potter and the Cursed Child,* when the protagonists travel in time, for them "time stops. And it then turns over, thinks a bit, and begins spooling backwards..." Outside of fiction, even if it does sometimes seem to do strange things in our minds, time continues to move forward linearly.

So when cosmologist and astrobiologist Paul Davies insists that, "there's no directionality of time; there's directionality of the world in time,"[16] that is problematic indeed. Surely directionality of the world is the directionality of time. In which way would the world be directional while time is not? Paul Davies also talks about "states of the world," as if they have no connection to time. Yet time is imbued into the state of the world; there cannot be one without the other. Making a case within science of a world that can only go in one direction while time could go another, is contestable as to what science is and to what it refers. It's as if science can separate itself from nature through mathematical solutions, allowing theoretical behaviors that otherwise could not occur, but that are reasonable and logical in themselves and for themselves. We will come across this strange dichotomy again as we progress along our path.

Opposition to the 'arrow of time' comes not only from science, that says it isn't *necessarily* so, but also from those who object to the basic idea of time being linear. Instead, time is posited by a number of religions and faiths as being cyclical. The idea behind the circularity of life is that it does not end with death and that the end of life is the beginning of something else.

Perhaps it is the supposed rigidity of linear time that so offends the *circularists*. The stricture of straight lines seems to put people's backs up. But linearity is rigid only in the consistency of direction. It doesn't mean that surprises don't occur on the way or that events cannot circle round to occur again. Knowing that time moves us toward the future does not, of course, mean that we know what is going to happen. The butterfly effect—any small occurrence that can be the trigger to cause eventual huge effects elsewhere in the world—does not affect the linearity of time, even though results could be monumentally catastrophic. Blips, large or small, can cause hiccups; but these do not change the linear direction of time. Time moves forward, whatever we do and whatever happens. Neither the probable nor the improbable changes that.

Realizing the significance of 'time's arrow' on Earth not only facilitates physicists in their search toward understanding the deepest mysteries of the universe; for us it is a reality that allows us to examine how time works within our daily lives.

And for that, we can do no better than use the river as a descriptively fluid analogy from nature, traditionally taken as a way of examining the conduct of time. Already in around 500 BC, the Greek philosopher Heraclitus made the by now well-known statement that "No man ever steps in the same river twice." Marcus Aurelius made a somewhat similar point some 700 years later: "Time is like a river made up of the events which happen, and a violent stream; for as soon as a thing has been seen, it is carried away, and another comes in its place, and this will be carried away too." And in Japan in the 12th century, Kamo no Chōmei, in the work Hōjōki, described the river as a metaphor for constant change:

> "Though the river's current never fails, the water passing, moment by moment, is never the same. Where the current pools, bubbles form on the surface, bursting and disappearing as others rise to replace them, none lasting long. In this world, people and their dwelling places are like that, always changing."

It's natural, this feeling that time flows along (and known just as commonly as 'the passage of time'). There is not, though, only one way to envisage what that means. Do we imagine time flowing around and past us (something like standing in the river), or as time sweeping us along (as if on a boat on that river)? With time flowing past us, we would be conscious of water hitting us from the past and then flowing onward toward a beckoning future; if, on the other hand, we were part of the flow, we would be in a perpetual 'now,' seeing the past and future only from where we always are.

Perhaps it makes no difference, just as when we are on a train at the station and don't know for a moment whether it is our train moving or the train next to us that is pulling out. It's actually two sides of the same coin. Since everything is moving relative to everything else, there is no such thing as objective motion—and hence no objective position—either in location or in time. Time flows whether we consider ourselves to be stationary, with time flowing past us, or whether we are part of the flow. To put it crudely and to the point: we get older whatever we do (or don't do).

And that is the most important point about time: nothing stays as it is, not even for a split second, and hence everything is always new. In fact, we *never* do the same thing twice or even see the same thing twice. Every event is unique, since it happens in a certain way only once in our lifetime. The date 12/12/12 was no more unique in that sense than any other date; each date occurs once only. Once, and that's it.

Knowing about the arbitrary nature of dates, though, does not change our involuntary reaction to them. We still get excited by 'unique' dates and most of us get a thrill from the arrival of New Year's Eve, standing, perhaps, in the freezing cold to watch fireworks, the event hinging entirely on the willingness of all participants to share and celebrate that one precise moment in time. We, as creatures of time, go along with it, even though we know that 'the moment' cannot be isolated from others within the continuum of time, the continually changing everything.

It's not only that the river is different every time we step in, but that even when we are standing in it, it is changing and we are changing. We don't need to do anything for change to take place; it just does, even though it would be nice to believe that we can remain unchanged and watch time flow past us. But like our view through the window of a train traveling through the countryside would give us a very real impression that it's the trees that are moving, everything, in fact, is moving in relation to everything else. There is no standing still: time won't let us.

Trains, rivers, movement, time. The river image is particularly apt as a way of illustrating the idea of the flow, or passage of time. But we can just as easily use other metaphors to show the constant change: we would also not be able to step twice into the same still lake or walk down the same road twice. Everything moves; nothing is entirely still, including ourselves. Time acts on us, whatever we do or don't do. For anyone who has read a book for a second time, it is common to see completely different aspects within the book, as if reading a different book. What changed? Of course the reader changed. It would be easy to say that the book hadn't changed; but that isn't strictly true. The words changed in the reader's eyes because of how he saw and related to them a second time. As actress Chloe Pirrie, who played Emily Brontë in BBC's drama *To Walk Invisible: The Brontë Sisters*, put it: "I read Wuthering Heights as a teenager. As soon as I got the part, I read it again. It felt different. I was struck by the brutality of it."

Let's leave for a while the image of stepping into the river and standing there, and go for one in which we float along with it, as if on a boat. This analogy would seem a perfect fit for the Buddhist and New Age advocates of *now*. It reflects what they mean when they state that we are always in the *now* and that that is all there is, with no past and no future. *Now* is, as it were, the boat. As we float in the *now* we pass by events that we then leave behind while we continue always to be in the *now* boat. In that sense, the past is unreal, since it is no longer with us; and so is the future unreal, since it is not in the present, either. When Pema Chodron, an American Buddhist nun says, "wherever you go, there you are," she is referring to a perpetual *now* as well as a perpetual *here*, encompassing a feeling of comforting safeness. Unlike the Duke in Thurber's *The 13 Clocks*, who was afraid of *now*, New Agers perpetuate the safety of being where we are.

Flying in the face of physics that tells us that there is nothing fundamental about *the present*, and so is not 'more real' or 'special' than the past or the future, Alan Watts uses the analogy of a boat on the river to make the point that the past fades away and does not affect our present. His observation that "the wake doesn't drive the ship, anymore than the tail wags the dog,"[17] is echoed by Eckhart Tolle, who says that, "The past has no power over the present moment."

Of course, Watts is right that it is the boat that causes the wake, not the other way round, just as a car causes exhaust fumes. Yet the analogy is misplaced; indeed, the tail does not wag the dog, but implying that the past does not affect people's present is far off the mark. Interestingly, Watts did not use the thrust of a jet engine as an example, since that *does* cause the airplane to be pushed forward.

Surely, though, the past is important to what comes after. In many ways it does drive what we do in the present. What could we do if we had no past experiences on which to lean? An extreme example of what that could entail is the case of Henry Molaison, whose memory was completely deleted in 1957 owing to a disastrous brain operation that was supposed to cure his epilepsy. He was left in a permanent state of amnesia, unable to retain any new information. Until he died in 2008, Henry was stuck in a bordered present and was unable to function properly: he did not remember what had happened a moment previously and so not recalling even the people closest to him, he had to be redirected to his room every time, and generally was lost in his world of the present. It seems that the present without the past is barren and quite unlivable or, as Cormac McCarthy said in *All the Pretty Horses*, "Scars have the strange power to remind us that our past is real."

We cannot move forward without memory. Knowing where we are includes knowing where we were; without knowing what we are doing here, we are lost. People with Alzheimer's and dementia are 'not with it,' because they have lost their past: their *now* moves back into oblivion, and so they have no memories to move them forward. It is another way of looking at the arrow of time, inevitably directing the past toward the future.

It's not only conscious memory of past events that moves us forward. Everything we do is premised on what has come before and that it will continue in the future. We get up in the morning with the understanding that it will be light during the day, that our clothes will be where we left them, that gravity will still cause objects to fall downward, that it will get dark at roughly the same time as yesterday, that cats will not suddenly fly, that people around us will speak the same language as they did yesterday and a host of events that we inevitably take for granted. In fact, our lives are based on what has come before and we act accordingly. In short, we accept the inevitability of the future based on the past.

On its own, *now* is like a single piece of a jigsaw puzzle, useless on its own without a greater context.

The case for the permanent *now* looks bleak, even if we generously extend it to encompass *the present,* particularly if we take into account our not knowing where it began and where it ends. Our boat is certainly pushed by something and is going somewhere.

And is it really *the present* that we are always and exclusively aware of? Imagine a walk on the beach, our feet on the sand, looking at the shells lying around. Without thinking of it consciously, we are aware that those shells once contained living organisms, that the waves on the ocean are part of past waves and future ones, that we are as much *was* and *will be* as we feel we are *is*.

There is another problem—so many problems and so little time!—with the notion that only *the present* exists. It shows up when we see a picture of someone who is no longer alive. Surely that person existed in the past and we are aware of that fact. Similarly, when we watch movies that feature long-deceased actors, we acknowledge a time when those actors were alive.

And since we are mentioning some of the problems connected with *the present,* let's add another one. Referring to everything as occurring *now* could lead us to the unfortunate conclusion that *what might be* is as valid as *what is* or *what was*. So if I think about wanting to go on a trek in Peru and then for some reason think about my past trip to Spain, both those thoughts apparently occur *now*; yet, one event actually happened and one didn't. What value does 'the present' have in cases such as this? The notion that we are 'living in the now,' and particularly that that is all there is, is so restrictive as to have no connection to what may have happened. Living in the *now* disconnected from the past has a huge price tag: the difficulty of differentiating between what happened, what might have happened or what hopefully might happen.

So while it is simple and somehow naively refreshing to float along on an everlasting *now*, how realistic are these lyrics of Willie Nelson in the song 'Always Now'?

It's always now
And nothing ever goes away
Everything is here to stay
And it's always now

It's always now
But never was a used to be
Everything is still with me
And it's always now

So brace your heart
Save yourself some sanity
It's more than just a memory
And it's always now

Willie Nelson, *Act Five Music, Warner-Tamerlane Publishing Corp.*

But we refer also to time passing us by, the feeling that we are on the riverbank watching the river flow along and we not in it. It happens often, that feeling of 'not taking advantage' of an event, allowing it to flitter away.

And yet, aware as we are that time is finite, that our river has an end, we realize somewhat uncomfortably that what stretches behind us is always getting longer, while what is ahead of us is ever less. As the Beatles put it: "You and I have memories longer than the road that stretches out ahead,"[18] with the punk rock band Forgetters exclaiming rather more bleakly, in their song 'Die by Your Own Hand,' that "Clocks tick louder every year, calling us home."

If, however, 'now' is all there is, thoughts of personal time coming to an end are superfluous. It allows *nowists* to close an eye to the unfortunate fact that in the finite length of time from birth to death, *the present* would have to encompass an ever-shorter future and also a continuously expanding past!

But with *now* as a non-event that we never actually alight on because it instantaneously moves back to the past, our expanding past and contracting future would be a natural consequence.

The unreal present

We all carry around certain intuitive feelings, and naturally we are disconcerted when asked to question them. After all, what can be more intuitive than *the present*? The very thought that we are not living in *the present*, that, in fact, there is no such time as *the present*, goes against what we feel as an unstated truth. Yet we do have, without realizing it, ambivalent feelings about *the present*. How to account for the craze of photographing everything, that has become a mass addiction? It's an odd thing, this desire of people constantly to capture objects or events on camera even before—instead of, it seems—they have allowed themselves a chance to see them directly. Art galleries, that once did not allow the photographing of exhibits, now allow it, resulting in people snapping painting after painting, instead of looking at them 'live.' Made easy through the ubiquity of the smartphone camera, it's as if, unable to capture *the present*, people feel they can do so by putting it on hold for later, when it is the past. The fact that they will in all likelihood rarely look at the photographed works of art later means that the unreachable *now* has essentially been moved to *never*.

The present is not infrequently treated as a peculiar state that we find impossible to grasp. In some strange way, we often experience events that evoke feelings so powerful, so close and bright, that we need to move back in order to experience them. Perhaps that is why we often enjoy an event retroactively through photos more than when we were actually present. It may go some way in explaining why, when Olympic contenders march into the stadium at the opening ceremony, many of them are busy videoing what they are actually going through. Already thinking of afterward, adding to future memories, they separate themselves from the occasion itself, from a present time that is already slipping away.

And what is the spectacle of a sea of smartphones videoing a pop concert, if not to distance the spectators from the actual event? It seems that nowadays one needs to view something virtually in order to enjoy it. In other words, only a vicarious presence is accepted as obtainable.

Interestingly, performers themselves have begun to object. From Kate Bush to Prince, the UK band The Savages, Roger Daltrey of The Who and Adele, all have expressed frustration at fans who insist on digitally capturing a performance. As Daltry put it: "I feel sorry for them. Living a life through a screen, not being in the moment totally. I find it weird. They're all mad."

In a poignant appeal to his fans after his first week playing Hamlet in London in August 2015, Benedict Cumberbatch asked them to remember his performance in their minds rather than on their cameras and phones. "I can't give you what I want to give you, which is a live performance that you will remember and hopefully in your minds and brains, whether it's good, bad or indifferent, rather than on your phones. So please don't."

Adele, in a concert on May 30, 2016, in Verona, Italy, put it more forcefully when she scolded a fan for setting up a camera and tripod, and filming her concert—instead of watching the live show. "Yeah, I want to tell that lady as well: can you stop filming me with a video camera because I'm really here in real life. You can enjoy it in real life rather than through a camera."

Preferring to capture a unique occurrence indirectly rather than experiencing it directly is a peculiarly human trait. When thousands of people turn up at Stonehenge for the summer solstice, that once-a-year natural spectacle, and view it through their smartphones, what does it mean? Are they snapping or videoing it in order to capture it for viewing afterward? Surely it is the experience of being there that brought them to the site. Without that experience, they could just as well watch a video of it at any time. Besides, will they really watch it afterward?

Something similar is afoot with modern parenting. Sir Anthony Seldon, the outgoing headmaster at Wellington College, Berkshire, UK says of parents, that "They spend all their time at the school play videoing rather than watching the performance."

Nobody is immune to it, even people who should know better. At the memorial service for Nelson Mandela in December 2013, Danish prime minister Helle Thorning-Schmid and US president Barack Obama took a selfie (with British prime minister Cameron jovially trying to get into the picture). It was inappropriate, but it was also endemic of the times, where capturing the moment as a surrogate for experiencing the event itself also brings to the fore the importance of 'me,' in this case over the person whose memorial service they were attending. The vicariousness of being once removed is as close as one can get to *now,* it seems.

A similar thought is behind this when a reporter asks Giuliano Vibilio, head concierge at The Gritti Palace in Venice, "I'd like to try something new here—what do you recommend?"

Giuliano replied: "Try walking around the city with a personal professional photographer. Having someone to document your visit properly will make it truly unforgettable." More than unforgettable, it is an urge somehow to make it more real after the event.

Historian Frank Shapiro told me that when he at last came face to face with the Bayeux tapestry in Normandy, he was only able to really grasp and appreciate it when he arrived back home and revisited it through photographs. I can testify to a somewhat similar experience when I toured the blissful scenery of Norway, coming over ridges to face the breathtaking fjords. As overcome with the beauty as I was, it was when I looked at the photos later that I was able to fully grasp and gasp at what I had seen at the time.

Perspective is really all we have; it is always of the past: a projection of what happened. It cannot be anything else; as noted by the novelist Simon Leys: "The novelist is the historian of the present." Similarly, visual art, which interprets and reflects real life, is perspective—of the artist and of the viewers. Often, it results in images that are more powerfully real than the original. Are flowers in real life as vivid as the ones in a painting? In some unfathomable way, reflected nature is often more 'real' than actual flowers in nature. "The fields are greener in their description, than in their actual greenness," as Raimund Gegorius put it in *Night Train to Lisbon*.[19]

Primarily, the hyperreality of art is due to the separation between art and the actual. Distanced by time and location from what 'is,' we are able to see what we imagine *the present* to be. And distanced physically, we see flowers in a painting from a different position than we would be looking at those flowers in nature.

Michael Frayn, in his book *The Human Touch*, mentions the model of a train that we look at in its every detail and only then appreciate it for what it is. He says, "We have a sense of reality more real than reality itself. We feel that we are looking at something that is at last graspable. Literally so. We could grasp it in the palm of our hand and for once actually take hold of the world that so persistently eludes us." But even then, we are looking at merely a representation of reality.

Circularity

A view of time as being *circular*—that life does not necessarily end—is inviting and comforting. Instead of the prospect of one's life ending, time as a cycle of events is life as a continuing path, possibly in a different form (such as through reincarnation) and/or somewhere else (paradise, hell, purgatory... and a return at some point). It's good for death (it's not really the end) and good for life (eternity in the here and now).

Circularity makes joyful sense for those who are averse to the very idea of straight lines, which are deemed to be unimaginative and boring and don't have the allure of seemingly natural curves. It is for this reason that the expression 'thinking in straight lines' is used as an under-the-belt insult and an underlying criticism of the role of rational thinking.

It's not surprising that the notion of linear time—the progression of time: past → present → future, from birth to death—carries with it a feeling of unease and an urge to revolt against it. It is so much more hopeful to believe in the ancient Eastern religious concept of reincarnation, that through a long process of rebirths, people can eventually reach spiritual perfection. The much-used term 'karma'—that which a person sows in this life, he will reap in the next life in his reincarnated state—is part of this fervent belief in eternity through reincarnation. Not surprisingly, 'karma' is used willy-nilly to mean an aura, a spiritual good luck charm ('good karma') for happiness and contentment, as opposed to the opposite ('bad karma').

The cyclic vs. linear argument is more than just a difference of opinion about time, or even about life and death. It conveys a basic difference about how one views the world and one's place within it. In many ways, it represents the divide between Eastern and Western thought.

The disagreements between East and West are not always gentle, often infused with righteous superiority. When circularity is associated with harmony, openness and 'thinking outside the box,' and linearity with rationality and plodding thought, the overriding generality is a given. It is not surprising, then, to see those feelings encapsulated in book titles, such as: *Auspicious Good Fortune: One Woman's Inspirational Journey from Western Disillusionment to Eastern Spiritual Fulfillment*.[20]

History is also brought into play within the idea that legitimacy of an idea is boosted by its longevity, the circle as representing the cycle of life having been used in Eastern mysticism for thousands of years, before the advent of the monotheistic religions. Circles in general have had a long run, in almost all cases representing the unbroken cycle of life-death-life. They have been used in Celtic rites, the 5000-year old Stonehenge circle, the Ancient Egyptian, Greek and Roman beliefs, and in Babylonian rites.

Today, the circle is no less prevalent as an important symbol, widely used in almost all parts of the world. We can see its significance in the Kabbalah of Judaism, as the Mandala used to raise consciousness in Hindu and Buddhist meditations, the circle of the Freemasons, the Tao yin-yang symbol, Tibetan prayer wheels, circles as part of tribal customs in the Americas and Australia, circles within the symbol for infinity, and countless more.

It goes without saying, then, that the circle as a symbol of 'wholeness' has unquestioned significance, not to say inherently special (even 'holy') properties.

This is all well and good, but why should it matter and what significance is there to the circle being part of an extremely long tradition as representing the circle of life? The fact that something has been done for thousands of years is, surely, not in itself a reason for it being correct, reasonable or right. Sun worship, for example, had a history at least as long as that of Buddhism. Human sacrifice was prevalent in various civilizations for over five thousand years. Length of practice is not necessarily, it seems, a killer argument for a particular philosophy or faith.

If we take away the issue of its longevity, we are entitled to be puzzled by the Eastern notion that a circle is symbolic for flexibility and openness, while linearity is considered to be inflexibly closed. Would it not be just as reasonable to view it the other way round? It is, after all, circles that are closed. Whole and closed. Yet, we have been inculcated with the idea that a circle is open, so that if accepted, linear event lines would be seen as closed.[21]

The reason for the anomaly is buried deeply in the mists of time, within the tenets of Eastern thought that has become so attractive in the West. It is literally a matter of life and death. If for Eastern faiths, life and death are circular without a beginning and without an end, for monotheistic religions there is linear progression from God creating life on Earth to death and the afterlife. The arrow of time is not a cool concept for those who believe in cycles. There are nuances, though. The notion of reincarnation taken aboard readily by millions of people in the West since the early 1960s is radically different from the reincarnation as believed in India, where the ultimate aim is eventually to reach the highest state of enlightenment and so get out of the continuous cycle of life-death-life. Not so New Agers, for whom circularity is a solution, a means of staying in the *now* as an eternity.

All this is not to say that there is no circularity as part of our daily lives. On the contrary, circularity plays its part as reruns of events, both natural and man-made, on the linear arrow of time. One of the most resonant events noticed by our ancestors was the returning sun each morning and its disappearance each night, the lengthening, shortening and lengthening sunlight, and the connection with growth within nature.

We humans feel secure in the knowledge that events will reoccur, that life follows a pattern of which we are a part. Time and religion have been co-structured, with times for prayer, for rituals, for thought and for contemplation. The book of Ecclesiastes contains probably the best-known ancient text that expresses an intensified interest in the existence and importance of recurring periods of time: "There is a season for every purpose under heaven: a time to be born, a time to die… a time for peace and a time for war… A time to speak and a time for silence."[22]

An integral part of the three monotheistic religions is a strict regime of particular prayers at certain times. Muslims pray five times a day, Jews gather to pray three times a day, and have a specific prayer when they rise in the morning and another one for when they go to sleep. Christian monasteries are tightly structured around prayers that begin with the time to rise and the time to go to bed (the English word for midday, 'Noon,' derives from the Latin name for the midday prayer, or None). The Jewish Sabbath is the most important day of the week and is strictly limited within exact times; interestingly, the Sabbath is religiously the most important day of the year, in part because it returns each week. And what could be more cyclic than our 7-day week, our 12-month year, and our national holidays and festivals?

All of this developed slowly from long before time was literally described and prescribed in stone all the way to the calendar, that structured lives institutionally. While the Roman calendar was originally determined by the cycles of the moon and the seasons of the agricultural year, it developed to also determine feast days and the birth of the emperor. In the Christian period, the calendar became an important instrument in marking religious festivals, from which there was no deviation.

Rituals do not need to be religious or spiritual, of course, to be what they are. Personal rituals, such as daily exercise, types of foods and eating times are a part of our lives to a greater or lesser extent, with certain rituals actually helping in enhancing the perception of physical qualities. Recent studies have shown[23] that rituals such as saying grace, making a birthday wish while blowing out candles or the ceremonial cutting of the cake before eating can enhance the flavor of the food. The same study showed that a delay between a ritual and the opportunity to consume heightens enjoyment—literally a mouthwatering experience—for example, the elaborate rituals of the waiter opening the wine bottle and pouring a few drops for someone to taste. Or the not uncommon habit of separating one's food on the plate and leaving the best till last. Or watching the waiter cut and debone a fish at a restaurant, heightening the sense of taste through anticipation. Stretching out time as part of anticipation and foreplay is part of the enhanced joy of sex.

Cycles are, in fact, very much part of Western civilization. Cycles in fashion, art, music, food, entertainment... Things that are 'cool' today are reintroductions of what was 'hip' yesteryear. Nightclubs became discotheques that became discos that become clubs that become nightclubs again. Even Las Vegas has its cycles that keep the cash flowing: from gambling halls to big-name shows to mega clubs in which the DJs are as famous and sought after as any big-name singer who keeps pulling in the people. If the roulette wheel is an apt metaphor for life, then Las Vegas is the instantly gratifying, gaudy proof.

It is quite amazing how much of our lives repeat themselves, and it is not surprising that modern literature and entertainment have latched onto this by maneuvering the text cyclically. In James Joyce's *Finnegan's Wake*, the last line of the book joins the first line, while in the movie *Memento* part of the plot runs backward in time while the parallel part runs forward. In two episodes of the TV comedy series *Seinfeld*, the whole story runs backward. F. Scott Fitzgerald's short story *The Curious Case of*

Benjamin Button, written in 1922, which became a movie starring Brad Pitt in 2008, shows the devastating effects of being born old and dying young, with love frustrated and Benjamin Button's final days of early youth bewildered by a lifetime's experience.

All literature from time immemorial consists of variations based on a limited number of themes. Money, sex and power are the bases of all novels, movies and plays, with time and distance as the cyclic common denominator, put so well by Toby Litt: "The basic plots of Western Literature depend on separation by distance—Odysseus separated from Penelope; the Odyssey doesn't exist if Odysseus can catch an easyJet flight home, or text Penelope's smartphone. Joyce's Ulysses doesn't exist if Bloom can do his day's business from a laptop in a Temple Bar coffee shop."[24]

Distance and time as inseparable elements recur as the bedrock within literature as a reflection of life that inexorably runs forward. Unlike that linear path, that is continuously lengthened, cyclic events can have opposite points that end up close to each other. "We are not so very different, you and I," says Karla to his nemesis Smiley in *Tinker Tailor Soldier Spy*. This should not be surprising, considering that the circle is a closed whole. "You complete me," says the Joker to Batman in *The Dark Knight*.

Circular time could be described as subjective and local, akin to the wheels that go round and round, moving a vehicle along objectively general linear time. As much as it is a matter of contrasting forces operating in different directions to produce movement, it shows how circularity is essential within a forward, linear movement.

Stretching time

As impossible as it would sensibly seem, time can stretch for us. We all know the feeling. No matter how busy we are, we invariably are able to add a little more, almost always succeeding in getting everything done, filling the amount of time available. Time really does seem to be magically and weirdly elastic, somewhat like the air in an unburstable balloon, always and necessarily filling the available space.

In one way or another, what we are all aware of is that time sometimes seems to go faster and other times slower—and it does what it does often completely out of sync with our wishes. Annoyingly, it sometimes goes slowly when we want it to go fast, and fast when we want it to go slowly. Henry Van Dyke, whom we met earlier ("Time is eternal for those in love"), said it well: "Time is too slow for those who wait, too swift for those who fear, too long for those who grieve, too short for those who rejoice." Even more frustrating is that while time does its own thing when we are unaware of it, time slows down as soon as we think about it, even when we are enjoying ourselves, thus negating our enjoyment.

And here we come across a very human paradoxical feeling: as much as we would like our lives to slow down in the long run, we are very happy for time to go faster in uncomfortable, boring or painful situations. It's akin to wishing the wheels to go faster while hoping to slow down the speed of the whole caboodle.

Actually, we don't really know what we want when it comes to time. In *Catch 22*, Dunbar, a character in the book, notices that time flies when he's having fun. It follows, as far as he is concerned, that to have a long life, you need to be bored for as much time as possible. Since he hates clay pigeon shooting, that is all he does with his free time, so ensuring that he has a 'long' (but awful) life, as opposed to a 'short' (but great) life.

Most of us don't work that way, since we are always in a hurry to be elsewhere or to do something else. Unlike Dunbar, we have an aversion to 'wasting time.' Standing in line, for whatever reason, is invariably considered a waste of time, as is waiting for almost anything. And since our annoyance is leveled at those we deem to be responsible for our long wait, it is in others' professional interest to relieve us of this feeling, either by getting us to believe that we are not waiting for long, or by having us not notice.

An example is the seemingly long wait in the airport arrivals hall for suitcases to appear on the carousel. It was noted by professionals in the field that even when the waiting time was shortened considerably, this made no difference to people's perception of it being a long wait. The solution? Simply to lengthen the walk from passport control to the carousel. Doing something is better than doing nothing, even though the overall length of time is the same. It's the same principle as one's preference to driving a longer route rather than the stop-start crawl through a traffic jam.

Our perception of how long time takes is very fickle. People get very impatient waiting for elevators, for example. It was found that they were considerably less impatient when there were mirrors in the lobby, since they had something to do: look at themselves.

And how many times do we gnash our teeth at yet again being in 'the wrong queue,' the one from which we stare helplessly at all the others that are moving faster than ours? Not only have we witnessed this phenomenon, but professionals have also noted our frustration—be it real or perceived—and have worked out solutions. Queuing theory (which can be traced back to 1919, long before there were supermarkets, airport check-ins and other similar modern delights) is a mathematical study of queues that predicts queue times and lengths, so that models can be constructed to shorten the time spent in queues. One example of alleviating the frustration of being in the wrong queue—and we almost never notice when we are in the 'right' queue!—has been to eradicate the system of separate queues. With the system of one queue for all, at the end of which the front person goes to the spot that has been freed, no one feels hard done by. A long, snaking queue might look bad to start with, but everyone is in the same boat. Not only is it possible to estimate, and show, how long it will take to get to the end, but often as well the queue moves past posters or leaflets that can be picked up and read, which also speeds up the waiting perception.

Does time really slow down when we are bored and go faster when we are enjoying ourselves? It certainly seems to, and we are often surprised at how fast time has passed, while on other occasions it seems only too true that 'time lies heavily on our hands,' as happens when time becomes a conscious part of what we are experiencing. Thinking about time seems to slow it down, particularly for depressed people, who will often use phrases such as 'time seems to drag' to describe their experiences and their life.

It is our behavior and reactions within time that determine its speed for us. It doesn't matter that the same amount of time, according to the clock, will have passed for a person having fun and for another person being bored. Twenty minutes will have passed both for the person having a tooth pulled as for the one outside reading a gripping novel. For Cinderella, time passed very quickly; not so for her ugly sisters, for whom time must have passed very slowly as they watched the girl dancing with the prince. It's not the clock that decides on the rate that time passes for us. If time being relative means anything at all, here is the proof in all its subjective glory.

The fact that perception of time is personal and subjective—the rate of the passage of time being in the eye of the beholder—is strange, since we cannot measure subjective time; it disappears as soon as it is held, in a way like looking at an image in the corner of one's eye. Subjective time, in fact, exists only in retrospect, never when we are actually aware of it, never 'now.'[25]

Yet, we live in a world where objective time (clock time) is what we need to live by. Clock time is consensus time, without which society couldn't function. We all synchronize our watches and calendars, and we do so according to consensus time.

We take consensus time seriously because it is conscious time, within which, and according to, everything around us works. It is 'objective' only because it has been set by the authorities, and it seems natural, being more or less according to the position of the sun. And we go by it, despite the vagarious reasons behind its set-up. Apart from summer time (aka daylight saving time), instituted in order to give us another hour of daylight, clocks are changed for reasons that have as much to do with politics as for any other reason.

How else are we to understand the 15-minute time difference between India and Nepal! Or the Navajo Nation in Arizona observing summer daylight saving time while the state does not—and the Hopi Reservation within the Navajo Nation not keeping daylight saving time in line with the state! And there is only one time zone for the whole of India, as there is in China. With India's western and eastern borders some 2000 km (1250 miles) apart, India could well have three time zones. China chose to have one time zone in 1949—Beijing time to enhance 'national unity'—abolishing five time zones, despite 3000 miles from east to west, which presents some huge problems of communication and the peculiar situation of people in Urumqi, Xinjiang's capital, enjoying a sunset at midnight. And there are the Channel Isles, very much closer to France than to the UK (with the island of Alderney only 15 miles from France and 60 miles from the UK), having UK time, not European time. And while British cities kept their own time until the mid 19th century, which meant that Bristol and London were five minutes apart, we can put it down to the foibles of history. Ever so slightly crazy nowadays is New York City, that, because of a city ordinance specifying the last Sunday in April as the moment to set the clocks forward and the last Sunday in October to move them back, for 8 weeks a year is an hour different from the rest of the USA. The fact that the rest of the country standardized the changes as March and November respectively does not make much of an impression on the councilors of the city, who seem to relish living in the past for a few days each year. Not to mention different parts of Australia, that have half-hour time differences, and countries from Venezuela to Iran, Iraq and a slew of others, that have their own time for reasons too disparate to mention.

Being pawns in the decisions of the authorities, we can but wonder at the power of those who can decree the workings of time. It helps to be royal, of course, particularly a British royal. When he was Prince of Wales, the future King Edward VII had the clocks in Sandringham House moved half an hour forward in order to give an extra half an hour of daylight for hunting. This was later carried out also in the other royal houses, and was continued by King George V throughout his reign and a particular private reign of terror regarding punctuality.

But it is the computer that needs to be satisfied above all; witching hour is not midnight when summer time changes to winter time and vice versa; it is 2.00 a.m. for the former and 3.00 a.m. for the latter. Changing clocks from summer time at midnight would cause havoc, since it would become 11.00 p.m., and then at midnight the time and date would change again.

Whatever consensus time is—perhaps in spite of it—individually we often live in our own time space, where time seems to go slower or faster depending on what we are doing or how we are feeling. And it invariably seems to take less time to come back than to go somewhere, especially when it is somewhere new. Yet, however slow or fast time seems to have gone for different people, objective, consensus time will dictate what we will follow. In fact, the world would figuratively collapse if there were no consensus time, unconnected as to how time is perceived to pass by individuals.

Consensus time (the time of our clocks) and subjective time (our internal time—how time passes for us) clash constantly. Often we arrive late for an event because we were engrossed in something; objective time and our subjective time were not running at the same speed. How many of us are unfamiliar with the last-minute dash to be somewhere after having had many hours to kill beforehand?

Being aware of the difference between the internal, subjective clock and the external, objective clock is what keeps us functioning within the social milieu. It's not that subjective time 'fools' us; it obviously has a psychological function as a counterweight to external constraints—those 'demands on our time'—that allows us to be ourselves within ourselves. Keeping that balance is not easy; for many people, it's a continual struggle; although interestingly, as we have seen, depressed people are much more consciously aware of consensus time, showing a greater sense of reality than their non-depressed counterparts.

Yet, in many ways the gap between objective and subjective time is being eroded. The demands on our time are so great nowadays that our subjective time seems in danger of being eradicated altogether within the relentless march toward having to do more and more, faster and faster.

Moving us ever forward is the rapid advance in communication ability, although there seems to be a chicken-and-egg situation, in that one can't really say what came first—the advances or the urge to communicate faster. "If you build it, they will come," says Kevin Costner's character Ray Kinsella in the movie *Field of Dreams*. Once the possibility of faster communication is made available, a subsequent question is whether or not the ability to do more in the same amount of objective time means that our minds are working faster. Perhaps that is what thinking faster means: filling our objective time with more tasks and information.

Psychologists have, in fact, investigated whether we can speed up our minds within a constant, objective, consensus time. If that were so, we would become mentally more productive, get more high-quality decision-making done per fixed amount of consensus time and learn more per minute. Although this does feel to be the case, there is no way to measure it scientifically.

There is no doubt that time seems to slow down in certain circumstances. Something along those lines may be behind the oft-repeated statement that the very best sportsmen and women seem to have more time when they need it—a phenomenon attested to by top athletes, such as tennis players at the point of taking a crucial shot in tennis, or Grand Prix drivers jostling for position on the track, who report a sense of 'slo-mo.' Neuroscience researchers at University College London suggest that at a critical moment, there is a change in the way the brain processes information.

If there is such an ability to stretch time under extreme circumstances, this is an astounding phenomenon, beyond a purely psychological effect. It would show a possible evolutionary adjustment to the ever-increasing amount of information that the mind needs to assimilate nowadays. In sports as in daily life, we are witnesses to possibly a new human adaptation to changing surroundings.

Why is it that a glass falling from the table seems to do so in slow-motion mode till it smashes on the floor? Why do earthquakes feel longer than they are? And why does the 'return trip' always feel shorter?

Much of why internal time and consensus time seem to pass at different rates is to do with how much information we are taking in. Information and experience stretches time for us. A child's day from 9.00 a.m. to 3.30 p.m. feels something like what a 20-hour day would for an adult. Children experience many new things every day, so that while immediate time can pass quickly when they are not bored, their minds assimilate and store information that needs to be processed, stretching the perceived overall time; but as people get older they have fewer new experiences and time wraps itself comfortably round what it has already come across.

Perception of the speed of time is, then, largely a function of memory. Information that is not new is recognized and so we passed over it quickly without it being added to our memory.

Some of this may account for the common feeling that a return trip takes shorter than the same outward journey. The point is, though, that it is not the same journey, because the way back is more familiar, having more recognized landmarks in our memory that we didn't have originally. In other words, on the outward journey we take in new information. We go out as children and come back as adults!

Because this strange difference in our internal, subjective time between outward and return journeys is so common, it was inevitable that somebody would research this phenomenon. Niels van de Ven, a psychologist at Tilburg University in the Netherlands, had two groups of cyclists take the same route to a luna park, with one group taking the same route back and the other another route of the same distance. Both groups—not only the one that was retracing its steps—felt that the way back was shorter, showing, he says, that the familiarity explanation was incorrect. Instead, Van de Ven and his co-researcher, Michael Roy, thought that the results were owing to the cyclists' expectation of the return journey being shorter.

Interesting things happen with lengths of time. If we know that our morning run will last for one hour, that hour seems to take very much longer than the last hour of a 3-hour run. After 2 hours of our 3-hour run, we have 'only one hour' left. Our expectant memory changes our perspective of time, so shortening it when appropriate. There is another way of looking at it: when an hour is the total of our run, that hour is 100 percent, but it is only 33 percent of a 3-hour run. We can apply this to the 'return trip effect' in general, where the return journey is 50 percent of what the total journey would be before we set out and so seems shorter. So getting from 'here' to 'there' is longer than returning from 'there' to 'here.'

Something similar occurs in the well-known holiday phenomenon of the first few days taking ages and then speeding up into a veritable gallop toward the end. The beginning of the holiday, while we get used to it, is equivalent to the outer journey; from then on, the familiarity of being away causes the remaining period to speed up as it goes over what the mind feels is known. Not to mention (with or without applying percentages) that what remains of our holiday is getting shorter while what is behind us is getting longer, so that our backward-moving *now* is going over more ground already covered.

Is it not a truism, as already noted, that the years go by faster the older one becomes? When young, a year seems to take forever, yet when one is older, years flash by at a bizarrely dizzying speed. It makes sense in the light of our foray into percentages. One year for a child of 10 is 10 percent of his life, whereas for a person of 60, one year is only 1.6 percent of that person's life. No wonder that time is a veritable flash for the elderly. Not to mention, of course, is their realization that what is ahead is a far smaller percentage than what has already gone—and getting smaller by the minute!

> Ticking away the moments that make up a dull day
> You fritter and waste the hours in an offhand way.
> ...
> You are young and life is long and there is time to kill today
> And then one day you find ten years have got behind you.
> ...
> Every year is getting shorter; never seem to find the time...

'Time' from *The Dark Side of the Moon:* Pink Floyd. 'Time' lyrics © T.R.O. INC.

Yet, the same cause that affects how time seems to fly for the elderly—the amount of information stored in the brain—is also why it takes them longer than youngsters to recall things. Their brain takes longer to work because they simply know more, much like a computer with a full hard drive that takes longer to access information.

Of course, one doesn't need to be old to be completely off when it comes to consensus time. Sufferers of insomnia are particularly bad at estimating how long they lie awake at night. This refers to about a third of adults, who every night have problems falling or staying asleep that aren't related to a persistent sleep disorder. And these sufferers can be extremely vague, a result of the difficulty that we have with accurately noting how many hours we truly spend sleeping on any given night. Patients who have spent a night in a sleep lab, for instance, often complain that it took them more than an hour to fall asleep, while a chart of their brain waves shows they were asleep within 10 minutes. Problems of self-reporting aren't limited to judging how long it took to get to sleep. Some patients wake up claiming that they hadn't slept at all during the night, despite hours of video and brain wave evidence to the contrary.

It is a part of the paradox that sleep presents to a conscious mind. Sleep is the absence of consciousness, which is impossible for the person to monitor. Since we cannot measure the time we are unconscious, what we do measure is the time we are awake when we know we should be asleep. The experiences of looking at the clock again and again, turning the pillow over hoping that the other side is cooler, turning from side to side looking for the magical sleep spot, all these seem to take ages, infinitely longer in our minds than the time we spent sleeping, which we simply don't remember. For us, of course time does not pass when we are in a dreamless sleep. We can only know it had done so when we wake up. "Only that day dawns to which we are awake," writes Henry David Thoreau in *Walden*. In other words, there is no dawn if you don't awake. For others there is, but not for you.

Being asleep is always in the past. So did time pass when

we were asleep? Of course it did—for everyone else. Not for us, though. And we know it did for us, in retrospect. All we need to do is to look at what our clocks tell us for us to know that consensus time had moved forward, the outward indication of our having aged. People in comas age, too.

All of this brings us back to the presupposition that there is no such a thing as time without change. For the person who is asleep or in a coma, time does not pass. Nothing passes for the person who is unconscious. Time is relevant only for a conscious person.

All this is for a totally dreamless sleep, in which the person isn't aware of being asleep. But if that person dreams, things are different. Time does not stop then, for dreams give the person a sense of time passing, even if within a weird non-realistic dream scenario.

Yet in certain circumstances, perhaps in meditation, or in a drug-induced state, time can seem to be standing still. It doesn't, of course. However wonderful the state of stillness and completeness, and however real it feels, the person is getting older. Consensus time does pass, and it does so despite the subjective feeling of a timeless present. Gurus and holy men in caves grow old, too. The 16th-century mystical poet Angelus Silesius put it this way:

> Time is of your own making
> Its clock ticks in your head.
> The moment you stop thought
> Time too stops dead.

Time, of course, could not stop. We cannot even imagine if it did, because we wouldn't be around. Birds and airplanes would stop in mid-flight, as well as ships, cars and planets in mid-orbit, and subatomic particles would cease moving. The hypothetical scenario would have birds and airplanes *not* dropping from the sky, the Earth ceasing to spin, our hearts stopping, as would all bodily functions. Time, in other words, is everything.

Getting through

Some 500 years separate the invention of the printing press and the Internet. While both changed the concept of time through breakthroughs in how information was disseminated, the Internet is unparalleled not only in the sheer numbers of people able to communicate and to access information, but also, and particularly, in keeping us informed of what is happening while it is happening. The Internet brought time to our doorstep, so to speak. Instantaneous information. Or, rather, the illusion of instantaneousness, and a strong one at that, inundated as we are with news and information of what is happening at the moment. Never before have we felt so much part of a 'universal now' as provided through the well-worn concept of one global village.

It's a delusion, though. We aren't any closer to *now* than we ever were, despite the seeming headlong rush toward its embrace. With *now* always, immediately and continuously slipping to the past, we can't even claim that it is doing so faster than previously. There isn't anything faster than the unmeasurably instantaneous, which is how fast *now* has slipped to the past.

It threatens to be overwhelming, not only how fast information gets to us, but also the sheer mass of information that is available. The more knowledge and the faster we receive it, the less time we can devote to each item. The flow never seems to stop, somewhat reminiscent of Newman's rant in a *Seinfeld* episode about the mail: "It keeps coming and coming and coming. And there's never a let-up. It's relentless..."[26] If previously people needed to catch up on the news, they were well aware that they were looking back at what had happened. But nowadays, everything is happening 'live'—an illusion fostered by news channels bringing constant news flashes as well as streaming on social media within the unrelenting now, now, now flow.

Where we used to be satisfied with 'things taking their time,' we are now often in a position to decide ourselves how much we can cram in. Box set binges, the practice of viewing entire television series in one sofa-hugging splurge, is a good example of instant satisfaction in the extreme. Facilitated by VOD (Video On Demand) and online streaming, the immediate consumption of an entire series feeds into a modern idea that faster and quicker is better. It's as if time as a gap is seen as an obstacle in the way of our knowing what comes next.

From this viewpoint, what would be the point of waiting for a week to see the next episode of a series or, as was the norm in the 19th century, to wait for the next chapter of a Dickens novel published each week in a magazine?

It's a sign of the times, this urge for more and more, faster and faster, and more frequently, fueled by an all-pervasive 'now' atmosphere that tells us all is possible if we want it. 'Living the moment' while we try to hold onto it through selfies as the moments flash past us is all the rage as a sort of alchemist-type recipe for making life longer and happier through grabbing everything available.

Too much can be as bad as too little, though. Eventually, it may overwhelm us to the point of overload, where it becomes white noise, akin to feedback, in which the events and their circulation are merged so that the information simply disappears. French philosopher Jean Baudrillard refers to it as a short-circuit between cause and effect, where it is impossible to discern which is which.[27]

Whatever happens, though, time continues, so that everything is necessarily transient. It's what time does: keeps everything temporary as we are swept along.

It took a while for technology to reflect the change from a more static viewpoint of the world to the reality of fluidity. It was a case of the technology being in place before we realized what it was actually doing. Taking the Internet as an example, the initial general perception of it was largely static, in that it took into account the notion of locality but not of time. So although it had moved our attention to computer screens, we were still acting in a static sense and in terms that reflected this: we were seeing things on a *desktop* in which we had *files* and *folders*, websites that contained *pages* and *platforms* that were ours when we downloaded them.

Bearing that in mind, 'cyberspace' is an anachronism, and should rather be called 'cybertime' or, even better, 'cybersphere.' Symbolically, this was reflected in the name change from what was Facebook's 'wall' to its 'timeline.'

It is not surprising that the shunting aside of static objects to accommodate the obsession with saving time would result in drastic changes. Not only the practical, but with it, also the cultural. The typewriter has slipped into the collective memory of museums, while the piano, once a part of the furniture in many houses, has moved aside in time and location.

As a personal possession, the piano is almost a thing of the past, attested to symbolically through Harrods of London announcing, in early February 2013, that it was closing its piano department, since it "no longer warranted the space." More than the implication of the Beatles' 'Move Over Beethoven' (and its now-quaint lyrics, "I'm gonna write a little letter, gonna mail it to my local DJ"), it is emblematically the instant satisfaction of listened-to music replacing made-music.

Time for art

"We're all individuals," the crowd chants after Brian in Monty Python's movie *The Life of Brian*. That humorous line encapsulates the anomaly of feeling that we are like everyone else, while at the same time knowing that we are each different. The lone voice of the person who then shouted out, "I'm not," hit the nail on the head.

Feelings aren't shared precisely; we cannot know if someone's feelings are the same as ours, however close we are with that person. Not only that, but feelings can intensify over time and space ('distance makes the heart grow fonder') or, on the contrary, can weaken over time ('out of sight, out of mind')—and these feelings aren't necessarily mutual when they occur. But whether they intensify or weaken, feelings are not static. They cannot be. They change simply through the passing of time. And time always passes.

The fact that our feelings never stay the same has tremendous repercussions on everything we do and perceive. It allows us to see in 'a new light' what we have seen a thousand times before. It is what allows us to see a painting or hear a piece of music again and again, and still enjoy it. It's not the familiarity alone that draws us, but a novelty that pops into our awareness.

What is it that draws crowds to the Rembrandts, the Rubens, the Michelangelos, the Titians, the Turners, the Van Goghs, and a host of favourite paintings again and again? Often, of course, it is the fame and the knowledge that these represent 'culture.' And, yes, it may be the iconic nature of the *Mona Lisa* that makes people want to see it 'live' in the Louvre even after having already done so on a previous occasion. Perhaps it even disappoints: if for the first time, seeing it 'now' may not satisfy the expectation built up in the past, the disappointment a second time may be of that 'now' not matching the excitement of a previous 'now.' Or, as Pascal Mercier says in *Night Train to Lisbon*: "A feeling is no longer the same when it comes the second time. It dies through the awareness of its return. We become tired and weary of our feelings when they come too often and last too long."

But it is more than that. A piece of art is actually not the same as it was when we saw it previously. Every time we look at a painting, we do so from a different perspective. In a way, it is a different *Mona Lisa* that we see a second time; the painting has changed for us because we have changed. Not only that, but we are never standing in exactly the same place we did before; so, without being aware of it, we are looking at it from a different vantage point. It was put succinctly in C. S. Lewis's *The Magician's Nephew*: "What you see and hear depends a good deal on where you are standing; it also depends on what sort of person you are."

Museums and art galleries are not, then, spaces in which the gathered exhibits are divorced from time and preserved artificially as if "outside of time and inaccessible to its ravages," as the French philosopher Michel Foucault put it.[28] Far from it, for time allows nothing to stand still—not for us and through us not museum exhibits.

But let's not get carried away. In all practical senses and in the way we *expect* to perceive it, the painting we see again is as it was the last time. Not necessarily, though, with locations. We are not surprised when we revisit, say, Venice, Italy after an absence of a few years, to find that our feelings about it are different. We are aware of Venice looking or feeling different and we may also be aware that we (and circumstances, who we are with, the season, weather) have changed since the last visit. Yet, we don't expect a painting of Venice by Canaletto to have changed from one viewing to the next. It has, though. It is not merely that physically, paint may fade over time. Just as Venice will have changed for us on a return visit, so will a painting of Venice have changed in the sense that we are looking at it a second time 'with different eyes,' not dissimilar to what occurs when two people read the same text and get different information from it. It is not just a painting of an object that changes, but the physical object itself—say, a china teacup—changes over time. It happens regularly, in art, in science and in life.

Let's stop for a moment and take a breath. All of this is very nice in theory, but could we be different than we were, say, ten minutes ago? On the surface, that doesn't make sense. Twenty years ago, yes. Ten years ago or even five years ago: that is reasonable. Definitely physically, since we can see the difference in photos from those periods. And very likely we have changed over the years in our outlook, our viewpoints, the way we think about things generally. Time has taken its toll over the years, we can acknowledge that. But that we are different to what we were an hour or ten minutes ago seems harder to swallow, as there is no empirical evidence to any difference.

Yet, it does make sense. As mentioned earlier, we don't notice changes as they happen, but only after they have happened. We don't see plants grow, just as we don't see children change. We notice how plants have grown and how people have changed when we haven't seen them for a while. Yet, if we haven't seen a young child for even a few weeks, we notice how much he or she has changed in that period. And how many times are we surprised at how a person we see on the television has changed in the five years since we last saw them.

All of this means that during the time we didn't see them, they—plants, children, adults—were changing. The change was so slow as to be imperceptible for those who saw them frequently; but it was happening. The inexorable march of time. (In our second chapter, when we look at memory, we will explore the reason why and how this happens).

If, then, there is a change from, say a year ago, there would be a change from six months ago or a month ago. Small changes, of course, but changes nonetheless. And if from a month ago, then from a week ago. Or a day ago. Although we don't notice the changes because they are so small, they must have been happening and are continuing to happen.

We are, of course, referring to changes so small that they are almost immeasurable, as are the differences in location that change the length of time it takes for each of us to see an object.

So, yes, when we see a painting after a week (or after five minutes), it is different for us, and it is so because *we* are different, albeit imperceptibly. William Faulkner got it only half right when he said, "The aim of every artist is to arrest motion, which is life, by artificial means and hold it fixed so that a hundred years later, when a stranger looks at it, it moves again since it is life."

It's just as well that we don't notice tiny changes where we don't expect them. We take it for granted that the people we work with will look as they did yesterday and that there will be no change to the chair we sat in yesterday. While we can acknowledge that things change over time, we do not expect changes to occur from moment to moment or from day to day. We function because we expect the familiar. Just imagine what a state we'd be in if objects changed every time we got back to them, as paintings and newspaper photos do for the characters in the J. K. Rowling Harry Potter series. In our world of reality, though, we cope with change that is expected, and are put out by changes that are unexpected and radical.

While we expect the expected, and expect time to tick along at a steady rate, we sometimes wish we could manipulate it. Often, we want time to slow down, for example, if we are at a superb concert or come across a work of art for the first time, as described by Michael Clune in his aptly titled book *The Quest for Permanent Novelty*. Lost in what we see, time seems to lose its normal flow in what seems to be a deliciously slow motion of feelings.

Time is slowed down perhaps by our taking note of things we wouldn't normally, more so when we notice details depicted in visual art—stones, trees, grass, the wrinkles in a face, the stitching in a hem, the shadows and more—or our puzzlement and effort to understand the lines and colors in conceptual art. It is paying attention to details in the way that children do but that adults take for granted.

Is that not part of what art is about? Is it not to look in a different way at what we otherwise take for granted in nature and our daily lives? Perhaps the extra vividness, the brighter colors we notice in art, mentioned earlier, hit us because we have become jaded and no longer notice them in nature.

Not only life-like paintings. When we ponder a work of modern art, it might make us see—literally and figuratively—what we hadn't seen before. As Edmund Burke, the Irish statesman and philosopher of the 18th century put it: "When we accustom our eyes to what is around us, they cease to affect us." He advocated allowing a fuzziness that would take us away from the naturalness and boredom of a painting, and get us to see it anew. This is precisely what modern art does, from Constable, Turner, Monet and onward to conceptual art, that gets us to see things in a new and different light. The notion that vagueness makes art clearer is interesting and applies to other fields, as we will see.

In the context of time, the dynamics are clear. For a start, modern art, especially in the geometric shapes within conceptual art, makes us stop and wonder, for what we have in front of us is a challenge and an invitation to look longer. But it's much more than that; it encapsulates the difference between conceptual art and classical art as a representation of what is familiar to us.

Conceptual art is timeless, whereas classical art is rooted within the familiar and often within a certain period. We cannot be anywhere else but in Venice when looking at a painting by Canaletto, while the beaches, meadows and Gothic cathedrals of Normandy are vividly brought to us through many of the Impressionist artists of the late 19th century, that include Paul Gauguin and Claude Monet. And while we can wonder at Picasso's cubist representations, they are the artist's interpretation of what we know in the world, be it *Woman seated at a window* or *Guernica* showing the horrors of war. This is, of course, very different from what goes on when we are faced with a work that may consist of a black square or seemingly random lines on a canvas, and which are not located within a particular place or time.

All of this brings us back to the thorny question of where we are and how time affects what we see. Faced with a familiar work of art, be it visual or a piece of music, time often seems to go faster, possibly because we don't dwell on details that we are already aware of. But not always. And that's the point. We can never know how we will react to art, since we are doing so in an amorphous perceived present time that moves into our memory and looks forward to what we expect; the experience is different each time because on one side (where we are coming from) we are affected by memories, while on the other side (where we are going) we are affected by our expectations that may be changing. It is not surprising, then, that our experience changes each time. This is nicely illustrated when we watch an episode of a comedy series that, despite having seen it many times, we enjoy again, because we are already anticipating and laughing at the remembered lines that are to come.

And that is what makes the perceived so slippery that we can never actually alight on it—it is part of the continuum of time, along which we are necessarily in a continuous state of movement and instability.

Art in time

During the first few pages of this book we noted that two things cannot happen at the same time. Nor can we look at two things at the same time. This is clearly illustrated as you run your eyes along the words you are reading in print or on a screen.

It's an illusion that we are seeing the whole painting, photograph or page of a book—or anything at all in its entirety, for that matter. We don't see the whole thing, since we can look at only one point at a time. Rather, in order to get an impression of ('see') the entire picture, our eyes move very rapidly from one point to another and back again. The illusion is fostered by images seen 'from the corner of our eye,' that makes us look again at what we hadn't seen properly. But often, when we look again, mirage-like it has disappeared.

So when do we see the whole picture? The answer is we don't, and certainly not in *the present* or any time that we can call *now*. We see bits of it and each bit at a different time, since we are not seeing them together. The *here and now* is, in fact, meaningless, since we are never wholly *here* or wholly *now*. The closer we get, we see even less of the whole. That's why it is recommended to stand back in order to get 'the full picture.' And if we actually see the whole picture, it is when we are so far away from it that it is a dot. Moving back, back, back, we could be frivolous and say that we see the whole picture when we are so far back that it has disappeared!

Yet, we are assailed with the importance of 'seeing the whole picture'—a criticism that we have been looking at the details, rather than the sum total of them all. There is, though, something very wrong with the notion. Taken as a metaphor, it would perhaps not be out of place to call it false, even a dishonest one, since the basic assumption—that it is possible see the whole picture—is false. And even if the notion of 'the whole picture' is taken as meaning the most important facts about a situation, it leaves to someone else's judgement as to which details are important and which are the trivial facts that can be left out; it is, in fact, the opposite: taking into account the most important facts, whatever they are, would necessarily mean that one is *not* looking at the whole picture but, rather, parts of it.

The world as a seamless whole is an instinctively appealing idea. The constant exhortation to *see the whole picture* is in line with the modern emphasis on 'wholeness'—known more spectacularly as 'holism'—which seems to have been made into the panacea of many of our problems. The holistic approach refers to taking into consideration whole systems rather than the individual parts, its most popularly known application being holistic medicine.

The notion that everything is in some way interconnected and mutually dependent cannot seriously be disputed. Yet, from there to conclude that it is better not to deal with individual parts independently, there being a unifying element to be dealt with as the underlying problem of all the individual parts, is a large and unfounded supposition. Even if we accept the notion that a whole is the sum of its parts—not necessarily so: we may be more than the sum of our parts—does not always mean that the individual parts are not important in themselves and for themselves. Nevertheless, the efficacy of 'wholeness' has become such a truism that is simply not questioned.

Yet we shall question it, from different angles, later in this book, as we have with aspects of time and the all-pervasive, universal *now*. For there is a connection. The idea that a system—be it the human body, the world, the universe, time—is naturally balanced and necessarily harmonious should not be taken for granted.

It is a nice, comforting thought, that of harmony and 'universal consciousness,' and follows the notion that everything is part of something else. The problem here is that it is never-ending: every part is part of a whole that is part of a bigger whole, so that if, for example, climate control is dealt with only within the context of the Earth, would we not then need to face the fact that our planet is part of the universe and take that as the whole? As an ideal, it may be wonderful, but it could also result in solutions not being sought because of there always being a bigger whole: a case of *reductio ad absurdum*. This is the essence of holistic medicine, where ailments in a specific organ are not dealt with until the 'underlying problem,' i.e. the body as a whole, is dealt with.

But if, as we have opined, we can never see the whole picture—literally with art and figuratively with holism—we can perhaps do better by seeing parts of it. So it is with certain art that specifically benefits from our getting closer, when seeing more means seeing less of the whole. A prime example is Mark

Rothko, whose conceptual paintings consist of color forms that need to be seen up close, so that the closer one gets, the more one notices. It is a way of getting inside the paintings, in order to discover for ourselves the basic emotions that Rothko is trying to bring out. Yet, however well we succeed in capturing Rothko's basic emotions, these are not 'the whole picture,' but part of what that could possibly be.

Getting closer does, of course, mean that it takes less time for the image to get to our eyes and to our brain. We could say, paradoxically, that we are seeing more of less in order to understand more of a possible whole. Rothko tries to help us discover for ourselves what we see by not titling his paintings, but rather only according to numbers or colors—*Red on Maroon*, for example—and he refrains from explaining the meaning of his work.

Modern art has for some time understood that *seeing the whole picture* is an impossibility, and expands on that fact in order to make a point about art itself. German artist Tomma Abts is one of a number of modern artists who do not want art to reflect nature. Abts's purpose is to create a work that exists only in the realm of art, so she will interrupt or curtail any circle, line, arc or wave before it can develop into a form that resembles something in the natural world. Perhaps most of all, she wants to show that nature is not a perfect whole. Her works are seen best by getting close and seeing more of less.

The notion that we cannot see all of anything in one go is an ancient one, told already in Ancient India as a parable and became part of Jain, Buddhist, Hindu and Sufi lore. The parable is of blind men touching different parts of an elephant, with each of them giving a description of how the elephant would look based on what they had touched. The story continues with a sighted man telling the blind men that each have a part of the truth that taken together would give the true picture of the elephant.

The parable is flawed, though. The problem is that the man who could see is not much better off than the blind men, since he also sees the elephant only from where he is located, and so can never see the whole elephant, either. The parable didn't need the men to be blind. Even if all the men could see, the information that each gives will not together give a complete picture for the others, since each would see his piece in his own way because of their different locations.

The fact that objects and events are seen from different angles, literally and figuratively, so that we can never actually see anything in its entirety, should be worrying. After all, we all want to see the whole picture and generally we assume that we do.

There is, it seems, no position from which we would have an entire view of anything. As we move to bring hidden sides into view, the previously visible sides become hidden. We can't see both sides of a coin and all the way round, nor all parts of a 3D holographic object, wherever we stand.

At the time of Ancient Rome, the empire was so vast, with territory covering Europe, Africa and Asia, that each emperor could only know what was going on everywhere by sending emissaries out to report back to him. With the huge distances and the different times his messengers returned, the information received never gave a picture as to what was happening at the moment.

And so it is when we look at any object. Not only can we never see the whole of it, but because light hits our eyes at different times depending on where we are, what we can see comes to us at different times from different parts. If we add hearing and touch, which are processed at different speeds by different neural parts in our brain, the notion of 'the whole picture' is even more impossible.

This is demonstrated nicely when it comes to all forms of art. Art, among other things, reflects life. Paintings, which are two-dimensional, allow us to see only what is presented to us, the other side, of course, being the back of the canvas. No art, however conceived, be it three-dimensional, electronic or whatever, can show us the entire object, even if we agree on what the object is or what the artist is representing.

And we usually do not agree, sometimes not even as to what art is. An example is the work of the French-American artist Marcel Duchamp, who was one of the first to create conceptual art that shocked the art world (and everyone else) with his works that made ordinary objects—such as a urinal in 1917—into art, worthy of being exhibited in a museum. Duchamp was tolerant of any interpretation of his art, since he believed that any interpretation is as valid as any other; he regarded it as the creation of the person who formulated it, not as truth, which, he wanted to show, is in the eyes of the beholder. In this he was influenced by the writings of Henri Poincaré. Poincaré postulated that the laws believed to govern matter were created solely by the minds that 'understood' them and that no theory could be considered 'true.' "The things themselves are not what science can reach..., but only the relations between things. Outside of these relations there is no knowable reality."

Poincaré wrote that in 1902, but 2500 years earlier, Plato made the opposite point with his allegory of the cave (now known as Plato's Cave). The allegory describes a row of prisoners facing a wall that has shadows cast by objects being moved in front of a fire positioned behind them. The prisoners, chained from childhood, are unable to move their heads, so that all they can see are those shadows, while the sound of people behind them reverberates off the wall around them. The point made is that the images for the prisoners are real, but that in fact they are shadows of the real objects outside the cave, as is later discovered by one of the prisoners and then others taken out of the cave into the real world bathed in sun.

Looking at it from a wider perspective, Plato's image of the cave is a metaphor of human life: we are all prisoners in a cave, a world of concrete objects, seeing only shadows cast on the wall by light streaming in from outside. In other words, we think that what we see is real, that we know the truth; yet, everything that we see is a mere illusion, and that all we perceive are merely imperfect images of perfect ideas.

And like those in the cave, how do we know that what we see is real? We don't know. We can't know. We can only believe we know.

In the light of what we have looked at so far in this book, we would have to wonder about anything—whether material objects or ideas—being 'real' for everybody. We discovered that the notion of an objective 'real thing' is tricky, because it depends on our location and the time it takes for the light to hit our eyes (or for the sound to hit our ears).

But here come the questions of questions: Can something that is never seen, can never be seen and is, in fact, unknowable, be real in any way? Is there anything that is 'real' anywhere? In which sense is it real? These are the questions we raised about *now*. In which sense is *now* real if we never alight on it and can never experience it?

Reality, it seems, is not even in the eyes of the beholder—even if he feels it is.

Crossed paths

Distance and time are intimately linked, so that our perspective of one affects the other. This is despite the fact that we cannot refer to 'consensus distance' in the way we do to 'consensus time,' where the latter is convention according to national or regional edicts (summer and winter time, political and other reasons for the set time), while distance is by global agreed measurement (as, for example, the universally agreed distance between two points).

Yet despite the universally agreed distance measurements, in many ways distance is as approximate and flexible in our minds as is time, depending on circumstances. It's not that distances have physically decreased—15,200 km still separates the USA and Australia, as it did a hundred years ago. Psychologically, though, distance has a distinctly more fluidly relative feel about it than it used to. The feelings behind the contradictory sayings 'distance makes the heart grow fonder' and 'out of sight, out of mind' have been made somewhat less relevant owing to the technology that enables us to cover distance in a shorter time, and even more so by the ability to communicate face to face through a variety of messenger apps; 'out of sight' is fairly rare nowadays, no matter how far away we are.

Looking down from 35,000 feet through an airplane window over mountains—the Swiss Alps, say—on a clear night, we are likely to see small pinpricks of light emanating from towns and hamlets below. It takes but a few seconds to fly over each, passing over clusters of lights that reflect entire centers of life being lived below us, the inhabitants themselves being oblivious to our existence, to our speed and to our interest or indifference to what is happening to them from our vantage point. We are not 'passing through' as such. We are too far away and traveling too fast to experience a sense of touching anybody's lives. We are passing over. In the right mood, though, we can get ourselves to imagine what may be going on down there, lives lived, their worlds, stories, desires, dramas, births, deaths, marriages, tragedies and all that makes up human experience. Our passing over is a vicarious virtual passing through.

If we muse in that way about what is below us, what is it that we are doing? It is not *time* that is bringing us closer to the people below; on the contrary, *time* is taking us away from them as it always does. In fact, whatever is happening down there has already happened by the time we see the lights from up where we are. Down there, thousands of feet below, is far from us, in distance and in time, the lights from the houses taking a sliver of time to reach us. Yet, it is the speed with which we are passing overhead that encourages our feeling that we are close, that allows us to imagine the lives that seem so clear to us. To us in our plane it looks as though we are going over slowly (despite the ground speed), thus giving the illusion of closeness and allowing us time to contemplate the lives of those below. Or perhaps it is simply a case of seeing more when at a distance, similar to the experience of looking at art.

And on the ground? Consider what happens when we are on a train that flashes through a station without stopping, zapping through a town going about its business. What is an important *place* for people in that town is for us on the train simply an almost non-existing, undifferentiated *instant* along our path. For inhabitants of the town, on the other hand, our train passing through would mean nothing; for them, if they think about it at all, the train is a speeding inconvenience intersecting a path that is theirs. Instead of uniting people, paths can just as easily separate them.

The notion of crossed paths makes a convenient metaphor for both location and time, taken up with alacrity by New Agers, for whom everything has a reason and so crossed paths are never coincidences.

That is a rather audacious view. In fact, it is a breathtakingly anthropocentric view of time having a role and that the role is to serve mankind. From that point of view, there is no such thing as good or bad timing, since events happen at exactly the time they are supposed to happen—for a reason; for a personal reason, in fact. "I find you cross paths with people when you need them in your life… Our paths crossed when it was meant to be and not before," posts someone on an online website, while another poster tells us seriously that her best friend's experience was that "she crossed paths with someone to whom she was married in her past life and had to resolve issues with him in order to move on in this lifetime."[29]

All this is what is commonly known as fate, destiny, kismet. John Lennon summed it up most aptly when he said, "There's nowhere you can be that isn't where you're meant to be."

Yet, unless individuals believe that time works specifically only for them, the conclusion would have to be that time works its mysterious and fateful way for everyone within moves that are part of an elaborate, finely choreographed dance. Fate, then, means that whatever happens is predestined to happen. According to this, we don't really make decisions or have a choice, since if we did there would be more than one possible result; but that is stymied by the notion of fate and that paths are supposedly crossed because they are meant to be crossed at that point in time, not before and not after. After all, if "all paths are interwoven," according to Sonia Choquette, self-styled 'vibrational healer, and six-sensory spiritual teacher,' "we are connected to one another, and one person's misery eventually affects everyone else." If that is so, we are literally tied up within a net of interwoven actions in which we are domino pieces held up by the capriciousness of every other person and circumstance.

And, according to this view, it doesn't matter if fate seemed to have worked in an unfortunate way, for that, too, is a good sign that something positive will come of it. As Saroj Dubey of the New Age Foundation says: "Our daily lives are replete with examples of how what we perceive as catastrophic events or disasters eventually turn out to be the biggest blessing in retrospect."

This is a precept that New Age shares with religions, reflecting the need to make sense of a world that often cannot be explained by science or through logical thought. The idea that things happen for no apparent reason is deeply unsettling to many people. This is why people are always looking for signs; signs give meaning to what would otherwise be confusion. So if a person suddenly thinks about a bird, for example, and then sees it everywhere—in an advertisement, on the cover of a book, in their dream, as part of the name of a play, or many people with *bird* connections in their names, or even bird-shaped clouds—all this proves to that person that there are no coincidences, that there is order, that there is meaning and harmony in the world, and that humans have a part in it. There is another explanation: whatever sign we look for, we will find it—and we will find it everywhere as part of a self-fulfilling prophecy, which Karl Popper referred to as the rationality principle, according to which people act in ways they see as suitable to their purposes.

So if fate is aligned to how people act in order to make sense of their particular circumstances and that everyone affects everyone else, where does free will fit into the equation? The notion that there is no such a thing as a coincidence does not, surely, square with the New Age credo that we create our own reality and destiny. How, in other words, do destiny and free will live together?

According to Buddhism and New Age, though, there is no problem. Yet, we have an internally contradictory statement such as: "Although certain things may occur in one's life that may be destiny, we do have free will."[30] Or there is: "We are free to make choices but we are not free from the effects of those choices," which is from the Spiritual Encyclopedia, that then ties itself in knots by mentioning the "energy that we can use for a higher purpose or for a lower purpose." So it is maintained that we do have free will, despite everything happening for a purpose, the clincher being that "accepting responsibility for our choices and being grateful for freewill empowers us to make positive changes." The inherent contradiction between destiny and free will does not stop Vietnamese Buddhist monk Thich Nhat Nanh saying that "You can become a Buddha or you can choose to be an animal."

The notion *free will* gets even murkier when it is presented as the ultimate freedom of making choices by the same monk: "At any moment, you have a choice, that either leads you closer to your spirit or further away from it. Letting go gives us freedom, and freedom is the only condition for happiness. If, in our heart, we still cling to anything—anger, anxiety, or possessions—we cannot be free."

Whereas we may be entitled to puzzle over such contradictions, it seems that they are considered legitimate in Buddhism and New Age, following on the Tao notion that "If you understand, it means you don't understand," and Joseph Campbell's "If the path before you is clear, you're probably on someone else's."

What does all of this to do with time? Everything. If paths cross at exactly when they are supposed to, it means that time is there to serve the purposes of humans. It would explain telepathy as a meaningful phenomenon that occurs when mental paths cross—not as a coincidence, but because they were meant to cross at exactly the right time. This brings out another question: Who or what is causing time to be so accommodating? Does this not hint at a higher power? Time surely does not have a mind of its own.

The main objection to the inevitability of crossing paths (aka fate) is that it can never be proven to be incorrect, since we always look at it in hindsight. Even spectacular misses, such as Decca Record's rejection of the Beatles in 1962 ("We don't like their sound, and guitar music is on the way out."). Or the rejection by André Gide in 1912 of Marcel Proust's manuscript of *Swann's Way*, the first volume of *À la recherche du temps perdu*, can be put down to it all having meant to be. ("Turning down your book," wrote Gide afterwards, "remains one of the greatest regrets of my life.").

It can be assumed that New Agers will also read non-coincidental significance into the fact that those two occurrences happened exactly 50 years apart! Mind you, it also shows that having crossed paths once does not make one immune to making the same mistakes a second time. Ten years after rejecting Proust's manuscript, the same publisher (but not the same reader) rejected a manuscript by a young Irish author.

The "obscene" writing that was "blighted by a diabolical lack of talent" was by James Joyce and the novel was *Ulysses*. Just another crossed path that was meant to be, it seems.

Being here now, where we are supposedly meant to be, is a safe place. It is not surprising that it is held onto as a haven and as a situation that, when found, would give comfort and peace of mind. Paradoxically, as we have seen, being where we are supposedly meant to be would seem to result in total lack of control of our lives.

The question perhaps should be 'Where are we?' The answer for each of us would have to be 'I am here,' for we are where no one else is. If that doesn't sound exact enough, there is a reason: we are where we are relative to everyone and everything else. The fact that someone else will point to us and say that 'you are there' merely places us in relation to him. But the question itself is really 'When are we?', since the answer relates to what we are that changes constantly.

Rather than looking to 'now' and *the present* as the intuitive hook on which we can hang our wellbeing, it is the instantly moving *now* that affords the stability within the relativity of everything in nature.

II

Was

All passed

As intuitively strange as it is to claim that *the present* doesn't actually exist, it is nevertheless a reasonable conclusion, once we think of ourselves as being within a constant change to the past, continuously moving back, back into our memories.

It puts an interesting slant on the Buddhist concept: "No matter where you go, there you are."[31] Jon Kabat-Zinn, in his book of the same name, puts his case across by maintaining that "it could be either 'Wherever you go, there you are' or 'Wherever you go, there you are not.'" Prompting us toward the advantages of meditation and 'mindfulness' (a topic we will address later), he expects us to plump naturally for the former. But in the light of what we have discovered about our place in time, we do not come to the conclusion he expects: since we are never static in time, we are never exactly where we think we are. As crazy as it sounds when put this way, when all is said and done, it is more accurate to claim that "Wherever you go, there you are not!"

Time is weird. Or rather our place within time is weird, even as we accept the way it is. It's only when we think about it, do we stop and wonder. It is not easy to break away from what we feel is real, especially when that is accepted as 'common sense.' When Eckhart Tolle tells his interviewer that "past and future don't really exist," he explains it with a question and an answer: "Have you ever experienced the past? No, because everything you experience is always the present." Yet, presumably Tolle remembers having experienced the past, telling us often about how he *was* inspired by 'now.' It's called memory. "Being in the moment," he says, "frees us from past regrets and frantic future planning; it makes us appreciate all that is around us now."[32] Even if he is right about this good feeling, dismissing the past as a nonexistent influence even in memory is a tall order.

Not only New Agers. Even scientists aren't always able to free themselves from what they feel (and wish) to be real events, even when their own specific field of knowledge has proven something different. A good example is the neuroscientist.and well-known active atheist Sam Harris, who is prone to advocate, whenever he can and without the reasoning expected of a man of science, the notion of living in the present moment. "It is always now," he proclaims, even while admitting that "it's not quite true in the matter of physics," that now does not exist![33] So even while conceding that he is in danger of sounding trite and "being close to sounding like a Buddhist," he nevertheless, with ne'er a blink, goes on to extol the wonders of stopping one's thoughts in order to seize *now* as a way of getting over one's problems and pain. "There are ways to live in the present moment," he says. Amazingly for a scientist and philosopher, he declares: "It is the truth" (!) that we are always in the now. This to an audience of non-believers! It must make us wonder—no, we have a good idea—what his reaction would be to a believer in God stating "It is the truth," when proclaiming how wonderful he feels when letting his mind dwell on God!

Of course, we all feel as though we are living in a continuing present. The point is, though, that we aren't, as Sam Harris, when speaking as a neuroscientist, acknowledges by pointing out that we are actually not in sync with what we think is happening. Similar to our discussion in the previous chapter, he points out that inputs to the brain come at different times and that, as he puts it, "consciousness is built upon layers whose timing has to be different." Giving the example of your finger touching your nose, he states that it feels simultaneous, even though we know that it takes longer for input to get to the brain from your finger than it takes from your nose. "Our conscious awareness of the present moment," says Harris, "is, in some relevant sense, already a memory." When we feel it, it is already in the past. We are, in other words, always already the past and not 'in the *now*.' And this can be measured, according to a study by researchers from MIT and the University of California, Berkeley,[34] who noted that the present moment, as we are experiencing it this very instant, is a flashback to the previous 15 seconds of our life.

At this point, it might be tempting for *nowists* to again ask incredulously, 'Where are we living, if not in the present?' The answer is that what we genuinely feel to be *the present* is actually a continuously and inexorably moving past.

It is, surely, time to overturn the common perception that we are creatures with a past, who move through a present and have a future that makes available to us a series of possibilities. It is not that we are creatures *with* a past, but rather that we are creatures *in* a past that is constantly moving as part of what we think we are.

This is nicely demonstrated by something that may have happened to many of us: a great evening being ruined by something that occurred at the end, say an argument with someone, or falling down the stairs, or whatever. But how could an evening be ruined when it had, until that unfortunate moment, been wonderful? After all, you had enjoyed it until then.

It was, of course, the memory of the evening that had been affected, in fact changed. And it is often the last memory of an event or a person that sticks in your mind, largely displacing earlier memories. While everyone around us gets older, a person we haven't seen for a long time stays in our memory as he was when we last saw him; as soon as we see him again (and our possible shock at how he has changed), the old memory is 'corrected.'

Amazingly, then, it is not only that our perceived present moves instantly into a past, but also that our past is changed by what happens afterward. By the time we become aware of having an experience, the whole event has already become an account of 'what happened.'

Participants in the Olympic teams' march-past with their videos and selfies are desperately trying to hold onto a present that they know is already slipping into memory. And while we are at the Olympics, we can note that the 1972 massacre of Israeli athletes by Palestinian terrorists cast a backward pall over the entire Munich Games.

It is the memory closest to where we are that can affect the entire past experience, just as it is the last dream before we wake up that sticks in our memory. It is a speculative possibility that women would have only one child if the stark, real memory of the pain of childbirth stuck in their minds. But the result—the child—to a large extent and for the majority of mothers, overshadows and reforms the pain of giving birth. It is also the reason that the last speaker leaves an impression that can fuzz the impression of the previous speaker. In Shakespeare's *Julius Caesar*, not for nothing does Mark Antony speak after Brutus at Caesar's funeral oration; it is Mark Antony who turns a pro-Brutus crowd 180 degrees into a dangerous opposition.

The whole issue of experience and memory is problematic. After all, if what is happening is often influenced by the memory of the last event that happened—be it positive or negative—how can we be sure that what we experience is anything other than a changed memory? How, in fact, can we reconcile our experience with our memory? The economist and Nobel prizewinner Daniel Kahneman tried to do so by positing two selves within the same person—'an experiencing self' and 'a remembering self.'[35] As plausible as that may sound, though, it doesn't work in the light of what we have found within our discussion about time. As we have seen, any sort of 'experiencing self' would disappear immediately, its experiences slipping into the past instantly. Not only is there no length of time in which an 'experiencing self' could function, but also experiences or events would be retroactively changed by an incident occurring later. With experiences in the present being a possible illusion, any 'experiencing self' has the shelf life of a mirage. Experiences cannot be tapped; only memories of experiences can be. Daniel Kahneman, it seems, agrees that there is no such thing as an actual present experience, despite his 'two selves' position. "We don't choose between experiences," he says. "We choose between memories of experiences." Experiences are memories, in other words.

What about personal feelings? If experiences are constantly slipping into a past, how 'genuine' is an emotion such as happiness? If asked whether or not you are happy, that, actually, is a question about memory. And it makes no difference if you are asked 'Are you generally happy?' or 'Are you happy at this moment?' These are all questions about memory, whether very recent memory or more distant memory.

It makes us wonder about the repercussions in fields where personal feelings are gauged within a gamut of agendas. On small and large scales—from clubs to nations—questionnaires about contentment are important as assessments for future planning. On a sweepingly large scale, polls on happiness are quite the rage as an indication as to how countries fare along some sort of international happiness scale. Since the aim of many governments has been to create the greatest happiness for the greatest number of people, the measure of happiness is taken very seriously. A Gallup Poll in 2012 taken in 148 countries asked 1000 residents in each about their experiences the day before. Among other things, people were asked if they smiled a lot yesterday, if they felt respected all day, if they were well-rested and if they learned or accomplished something interesting.

The poll showed what were considered surprising results: those who live in robust economies like wealthy, business-oriented Singapore are among the least happy people on Earth, while the countries that reported the highest rates of 'yes' answers to these questions were Panama and Paraguay, with an 85 percent positive rate each.

Even without going into the various pitfalls of polls in general and the problem in defining 'happiness,' the whole basis for the poll is suspect—not to mention the attendant misleading headlines, such as "Happiest Nations on Earth Revealed."[36]

Whether the poll is about general contentment or the feeling of people at the moment they are being questioned, it is actually about memory. Even if they had been asked how they were feeling at that particular moment, it would really have been about how they had been feeling over a length of time, possibly influenced by what had happened to them recently. It wasn't even that, though: they were asked about how they had felt the day before, so even further distorted by what had happened during that time! The fact that these polls are often used as a guide on how to apply national economic policy, it should not be surprising that the results of these policies often turn out to be not as expected.

After all, we cannot separate the perceived present from memory. If all is connected to memory—and we are about to examine how reliable memories are—on what is there to lean when it comes to supposed present feelings?

Even the clearest experience of who we are turns out to be somewhat removed. The closest we come to seeing ourselves is our physical reflection, which, as already pointed out, comes back to us with a delay. We can, in fact, never see ourselves as we *are*, but only as we *were*. Any view of ourselves as we think we are is, then, necessarily distorted.

How strange it is that we don't know what we look like. Possibly there is an evolutionary and psychological advantage, allowing us to see ourselves as we would like to. In experiments[37] in which people were asked to identify pictures of themselves amid a lineup of distracter faces, participants identified their personal portraits significantly quicker when their faces were computer enhanced to be 20 percent more attractive.

How can we be so self-delusional when the truth (as other

people see it) stares back at us? Does Donald Trump not see how ridiculous his hair is? In the same vein, does the middle-aged woman wearing a miniskirt that is too high not see how inappropriately dressed she is? Not only do we not know how we look, but neither, obviously, do we know how others see us. 'How do I look?' we ask someone when we wear something new. It puts paid to the Socratic notion that we cannot know other people till we know ourselves. After all, if we can't get past a distorted view of ourselves, we have no chance of getting to know anyone else.

"Childhood is the time when we make grimaces in front of the mirror, middle age is the time when the damn mirror strikes back," says Bo Engström. But often we don't see it. Churchill loathed his official portrait painted by Graham Sutherland, and complained that it did not show him as he was and the institution he represented. To which Sutherland replied, "I can't be blamed for what is." What was for Sutherland certainly was not for Churchill, who had the painting destroyed.

Memories are made of this

Jane Austen declared in *Mansfield Park* that "If any one faculty of our nature may be called more wonderful than the rest, I do think it is memory." Memory is, indeed, an amazing function. We take it for granted, yet it is truly wondrous, this effortless ability to run at will whole scenes of our past through our minds, bringing up people both alive and dead, allowing them to behave and interact with us as clearly as if they were in front of our eyes. There they are, whenever we want them, within a whole library of past acts and events that we can access, and play back in glorious color and with accompanying sound. An anchor to what we are.

The claim that we are the sum of our experiences, our memories, makes sense. Whatever we are born with, that which is inscribed in our genes—our nature—is intertwined with the memory of our experiences.

Memories being an integral part of who we are, it seems reasonable to consider life as being made up of a string of remembered moments. It certainly does feel as though we move from one moment to the next, each one being complete before and after others, somewhat like a movie made up of a sequence of single frames. And like a photo, which reflects one instant that has passed and may never return, it does seem as though memory is a sequence of past events.

But just because time is continuous as is a movie, there is an important difference. Time, unlike a movie, cannot be stopped in order to look at a constituent part. There is no point of separation between what was past and what will be future within the continuum that is time, since we cannot isolate a section of our lives without taking into account the continuous changing past and potential future. Time is no more a sequence of 'past *nows*' than it is a sequence of 'present *nows*,' in the way that a movie is a sequence of stills. A memory is not a stopped moment, a frozen part of time that we can examine to see what really had been, even though it is tempting to think of it as such. Remembered moments have changed over time, so that what we remember today is different to what we remembered yesterday and will be different tomorrow. As time moves, so does memory.

If and when humans are ever cloned, the clones would be identical in all ways—but for only a little while; it wouldn't take long for them to differ mentally, owing to their inevitable varied experiences and consequent different memories. And you can bet that even the same event in which both were present would be remembered differently by each.[38]

It happens to us all the time when we come across a newspaper report of an event at which we were present that is not exactly as we recall it. Sometimes it is so different that we wonder if we were really at the same place at the same time as the reporter. Not only the media, of course. It occurs whenever someone recalls an event that we also remember, but differently. This is the so-called Rashomon Effect, after the 1950 Japanese movie of the same name, in which a crime witnessed by four individuals is described in four mutually contradictory ways. These different versions of the same event—also brought out in the 2015 television series *The Affair*—demonstrate how views are necessarily subjective, with different personalities bringing their own outlook to the equation, and also depicting how memories are genuinely different.

This should worry us. After all, however much we feel our memories to be genuine, we cannot, in all sincerity, believe that only our own memories are reliable—'true'—while everyone else's are not. The conclusion would have to be that memories are unreliable. That, in fact, all memories are. That leads us to a big question: Is there such a thing as a 'real' or 'true' memory? In fact, was there something that actually happened—even if no one can ever access/remember it? We'll get back to that question in various forms as we move along the twisted path of time.

That 'true memories' are unreachable shouldn't come as a surprise. In fact, it would be surprising if they were. After all, if everything that is happening is actually moving to the past, thus making our experiences moving memories, it would not be unreasonable for our memories to lose stuff along the way, while at the same time picking up bits and pieces from here, there and everywhere.

There are other points of view. For example, Canadian neurosurgeon Wilder Penfield claims that memory appears to contain, in great detail, every event that has occurred throughout a person's life and that individual memories are lost because there is no room in the brain for such detailed memory. Neurologist Steven Novella takes a similar view, pointing to memories being highly selective due to the brain having to choose what is needed from among huge amounts of surrounding detail. As Homer Simpson put it: "And how is education supposed to make me feel smarter? Besides, every time I learn something new, it pushes some old stuff out of my brain. Remember when I took that home winemaking course, and I forgot how to drive?"

From what we know, though, the notion that memory is infinite and that we remember what is useful in the present, proposed by French philosopher Henri Bergson, among others, is not a provable proposition. There is no way this can be shown to be true; after all, how could we access memories that we might have had? Not only that, but there are many times when an essential memory fails to materialize exactly when we need it. When British philosopher Galen Strawson says, "In spite of my poor memory, I have a perfectly respectable degree of knowledge of many of the events of my life,"[39] on what does he base that, and how would he know it to be true?

The increasing amount and speed of information that we have thrown at us results in our remembering less of it simply because there is so much, and not necessarily because some of it is not useful.

There was bound to be a backlash: disposable memories within social media sites, such as Snapchat ("Life's more fun when you live in the moment!" says its jolly slogan), that self-destruct a short while after photos and video are received, the idea being: snap it, share it, forget it.

It also helps that much of what is out there simply doesn't get into our memory. This is what the German sociologist Hartmut Rosa calls the 'television paradox,' that time flies while watching hours of fast-action movies, but that you remember very little afterward.[40]

So how can we possibly know that we keep memories in every detail, when our memories are often unclear and obviously cannot accurately reflect what happened? And how could we know why we have lost some of them? With all signs pointing to memories as being a very suspect record of what might have happened, there are many angles to the witticism that "Women worry about the things men forget. Men worry about the things women remember."

If we add to that the phenomenon of the constantly backward-moving 'present,' we realize that we cannot know how and when memories were formed, since this would have happened on the go. We cannot, therefore, assume that our memories were ever of 'the real thing,' despite the absolutely clear feeling that they are, since there was never any static *now* on which they could reliably be based.

It is here that we again arrive at a fundamental difference with the beliefs of Buddhists and New Agers, for whom the past is not real, with only *the present* existing. Following that line was the deeply pessimistic and nihilistic philosopher Arthur Schopenhauer, whose stark position was that reality can be only *the present*; as far as he was concerned, that which happened in the past is unreal—as unreal as if it had never happened: "What has been exists no more; and exists just as little as that which has never been."[41] But to maintain such a position—that the past once existed and it now *no longer* exists—would mean that the past did exist at some point. But the past was never *the present*: it was always the past. The past can only be compared to another point in the past within their continual backward-in-the-past movement. A 'present that was' is not an ex-present, in other words. That would account for memories of birthday parties when we were children that include our parents when we last saw them, rather than how they looked at the time.

Perhaps, though, it doesn't matter. After all, we know that there was yesterday, last month, last year, periods going back into our past; as far as we are concerned, our memories are of what really did happen. Yet, now and again, memories become vague or disappear for a while, then pop back when triggered by a smell, a sound, a word or a seemingly unconnected sign. As jolting as that 'involuntary memory' is,[42] it must make us wonder whether it is a real (unconnected to what happened in the meantime) past that we are thrown back to.

"When to the sessions of sweet silent thought / I summon up remembrance of things past," says Shakespeare.[43] Andy Warhol summoned back memories by sniffing at perfumes he had worn in the past—his personal 'scent museum'—so that he was instantly brought back to those periods.[44] It's a comforting thought that one can do that. But can we? Is an intense memory necessarily a real memory? Neurologist Oliver Sacks writes about events in his past that were etched in his mind as real, vivid memories; so real, in fact, that he says he could swear on them in a court of law. That is, until his brother informed him that a particular event mentioned by Oliver in a book, at which neither of them had been present, was, in fact, described to them afterward by an older brother.[45]

Memory gives you back what you think you had lost, sometimes giving you what you never actually had—a sort of wishful thinking and illusion of what could have been. In the television series *The Affair*, Alison describes that perfect moment when Noah pulled her to him. "He looked at me. At me!" she says. "I wish I could go back to that moment," she says wistfully. Her friend answers, "You never can." "Why not?" Alison asks. "Because it wasn't real," her friend answers.

Let's stop for a breather and clarification of an important point: the need to differentiate between *the experience of having a particular memory* and *the reality of what is being remembered.* Experiences are real. The experience of having what one believes is a memory is real. The question is regarding the reliability of the memory itself—whether or not the memory is of anything that happened and, if it did, how accurate the memory itself is.

The evidence against memories being of a 'true experience' is very strong. If, because of the nonexistent *now*, we don't know where and when we are, it follows that we can't be sure of where and when we were. "I am a very old man," Mark Twain said, "and have suffered a great many misfortunes, most of which never happened." Although not old, the actor Michael C. Hall pulled himself up over his memories. He mentioned that when he was in the third year of the TV series *Six Feet Under* he had what felt like real memories, which were actually the memory of a scene having been filmed two years before with his fellow actors. Memory was dismissed most starkly by writer Beryl Bainbridge, who said that "What we remember is probably fiction anyway."

It happens often (perhaps not often enough) that people are alert to the possibility of their memories not being their own. Speaking about his testimony before the Warren Commission investigating President Kennedy's assassination, detective Fred O'Sullivan, who had known Lee Oswald and had been in school with him, was questioned over a possible relationship between Oswald and David Ferrie, known to have dubious links to the mob. "I am trying to get things straight in my mind, just what I know and what I have heard," the young detective said. "It gets kind of confusing when you read so much. Sometimes you remember things that you don't really remember, you know."[46]

Memories do occupy a very shaky position. While 'now' is nowhere, memories, on the other hand, are everywhere, filling our thoughts as we go about our lives and being the base on which we plan our actions. Yet, while memories occupy such a central position, they themselves are a range of possible remembered events having happened in the past: memories of memories. And even then, they are battered and altered by what happens later. How can a memory be a totally accurate reflection of what happened, when what happens afterward retroactively changes what happened before? This was recognized also by the German philosopher Georg Wilhelm Friedrich Hegel in the 18th century, who said that change has the effect of altering the past, as well as the future. Remember the wonderful evening ruined by falling down the stairs on the way out?

It can hardly be otherwise that our memories are colored by what we are at the time we are looking back. What we felt and who we were then were different. Not only don't we know how we felt then and how it affected our experiences, we can't possibly know. All we can know is how we feel when we bring up a particular memory and how we remember our feelings to have been at the time. As Mountjoy in William Golding's *Free Fall* put it: "I am not a man who was a boy looking at a tree, I am a man who remembers being a boy looking at a tree."

Retrospective memory change is demonstrated in research studies, as in those done by Joe Voss at Northwestern University's Feinberg School of Medicine, Chicago, Illinois: "Our research shows that memories change over time to reframe and edit events to create a story that fits a person's current world," he says. Childhood memories are particularly fraught with the trap of today's viewpoint. We can't look back to our childhood other than with the baggage that we have accumulated since that time. Looking back from the standpoint of an adult, we don't see what we saw when we were children. As the author Anthony Rudolf says, "Childhood is an image, a collection of images, a construction, a fiction of reality…"[47]

How can we not be influenced by events that have occurred since? Shakespeare recognized this, and in Mark Antony's speech to the anti-Julius Caesar crowd he warns against the possibility of their memory being changed over time. "The evil that men do lives after them. The good is oft interred with their bones," he says.

The comment that spread widely on social media, "When the past calls, let it go to voice mail. Believe me, it has nothing new to say," is simply wrong. The past has plenty of new things to say if it changes the way we think today of what might have happened yesterday.

On the other hand, we could also claim that if we look at the past as if it were a road we have taken, looking back along that road will show us clearly where we have been and how we got to where we are. No, not really. That road, the one we see through memories, is not the same road; it will have changed in some way. Young Stephen Dadalus in *Ulysses* noted the problem when he said, "So in the future, the sister of the past, I may see myself as I sit here now but by reflection from that which then I shall be." We can't, in other words, see the road exactly as it was when we had come along it; our memories show it to us only afterward. Leaves lying on the road, cars that weren't there before, and we ourselves being different having gone along the road that we are looking back on. "Memory is deceptive because it is colored by today's events," said Einstein. Things simply aren't the same afterward as they were before. They can't be.

More than we realize, the past is subtly, and not so subtly, changed as soon as we know the outcome. Look at elections and how they are changed retroactively in our minds as soon as the results come out. No matter which contest, there are common features—the advisers to the winning candidates were geniuses with a ruthless but brilliant election master plan, while the campaign of the losers was one of chaos and bungles. If the same election had gone the other way, so would have been the brilliance of the campaign winners and the inefficiency of the losers. 'Facts' are—let's face it—merely how we understand them at the time we think of them. In some strange paradoxical way, we would have a clearer picture before the event, when the results are not yet known and we have not yet retroactively influenced our thoughts!

It can't be stopped, this continuous movement of a memory backward from when we are thinking about it. As the last line of *The Great Gatsby* has it: "So we beat on, boats against the current, borne back ceaselessly into the past." This backward movement means that the memories themselves are changing continuously. The more experiences we have—and these are going on all the time—the more our memories accumulate and distort themselves. And every time we access our memory, we are doing so from a different perspective, from a different time or mood, or perhaps dependent on how we got up that morning; any of these alter memories, so that the next time we access them will be based on, and compared to, the last perspective. The very act of accessing a memory will alter it, in other words. The unavoidable consequence is that the more often we remember an event, the more changed it becomes. Or as Alice in Wonderland put it, "I can't go back to yesterday because I was a different person then."

The old adage 'It's always darkest before the dawn' seems starkly true to us; we have all experienced something like it. Yet, it isn't any darker just before dawn than it was five minutes earlier, or, in fact, any time previously. Not at all. It's a retrospective illusion brought about by comparing the memory of darkness with the light of later, where the memory of darkness was altered through the influence of the light of the dawn.

We can see, then, that counter to what is most commonly taken for granted—that the past influences what happens later—it can justifiably be claimed to be the other way round, as well: what happens later can change what had been. It also means that we can never know 'what really was.' The notion that 'the present' can change the past is pretty mind-blowing and in that respect, it makes sense for actress Patricia Arquette to muse that "It's funny how you don't love someone until they're gone." The other side of this particular coin is Slavoj Žižek's contention that "When love goes away, you retroactively establish that it wasn't even true love. I don't only believe that I'm no longer in love," he continues, "I believe it never was." Whichever narrative you side with, they both point to a so-called present feeling changing one's past.

It is the retrospective change that accounts for why music that seemed so outlandish when first heard, in time becomes pleasant and mainstream. The composer John Zorn said about Thelonious Monk that "People used to think his playing was ugly. Now it's recognized as classic." Many of us don't remember, and it is hard to imagine today, that in the early 1960s the Beatles' music was considered by many as a cacophonous noise, as exemplified in *Newsweek*'s 1964 cover story on the arrival of the Beatles in the USA: "Musically they are a near-disaster: guitars and drums slamming out a merciless beat that does away with secondary rhythms, harmony and melody." It was acceptable in the movie *Goldfinger*, of the same year, for James Bond to compare drinking chilled champagne with listening to the Beatles without earmuffs. Today, it would be an incomprehensible sentiment.

Part of the process of getting used to new things is experiencing them over and over, with exposure having a lot to do with the popularity of certain works of art and, significantly, why particular paintings rise to the top of the culture ratings. Psychologist James Cutting, a professor at Cornell University, Ithaca, New York, calls it the "mere-exposure effect," which goes some way to explaining why, for example, the *Mona Lisa* is the most valuable and looked-at painting in the world with a constant crowd looking at it, when there are others by Leonardo da Vinci located very close in the Louvre, Paris that are passed by.

In experiments that Cutting carried out during his lectures at Cornell, undergraduates were shown works of impressionism for 2 seconds at a time, some of which were well-known while others were less well known but of comparable artistic quality. Students who had been shown the lesser well known works four times as often, preferred them to the canonical works, while a control group of students liked the canonical ones best. Cutting's students had grown to prefer those paintings simply because they had seen them more often.

The preferences were a function of memory, with each

occasion changing a previous memory. The more often a memory is accessed and changed, the fresher it is, compared to those memories that are accessed less often. So in Cutting's experiment, each time the unknown paintings were shown, they became more familiar and thus preferred to the canonical paintings that had become less familiar to the students. "The past," William Faulkner wrote in 1950, "is never dead. It's not even past." It isn't past in the sense that it is constantly in movement and change, restlessly being rewritten by what is happening afterward.

We should perhaps look afresh at the well-known lines from *The Rubáiyát of Omar Khayyám*, translated by Edward Fitzgerald, and wonder if that which is written cannot change in time:

> The Moving Finger writes; and, having writ,
> Moves on: nor all thy Piety nor Wit
> Shall lure it back to cancel half a Line,
> Nor all thy Tears wash out a Word of it.

Often, the past is changed deliberately through tampering with our memories. Political correctness is a stark example of deliberate semantic change in order to avoid offending sensibilities. An example is the strongly contentious word 'nigger' ('the N-word'!), which cannot nowadays be uttered or written—except by black people about other black people, as a term of affection—because it has become highly offensive.

Yet, retroactive banning of what is today offensive can be problematic. Deleting words that are now deemed offensive from the re-airing of movies and television series is a guileful attempt at changing the past, especially when it comes to words that when originally used were not considered offensive; censoring them today adds insult to injury by indicting as bigots those who had used them. And so a slew of TV programs are simply not re-aired at all, an example being *All in the Family*, where the whole point at the time was to show radically offensive behavior. Seemingly with the best of intentions, the Orwellian world is alive and well, changing the past to what it should have been according to the morals of today.

Whether the 'good old days' were really as good, free and innocent as we remember, we don't know. As nostalgically comfortable as it sounds, 'a trip down memory lane' is possibly a trip to a false past. As American philosopher John Dewey put it, "Time and memory are true artists; they remold reality nearer to the heart's desire." It's somewhere that we think we would like to be, usually better than what has happened later. Here is how James Joyce wondered about it in *Ulysses*: "I was happier then. Or was that I? Or am I now I? Can't bring back time. Like holding water in your hand. Would you go back to then? Just beginning then. Would you?"

Memory lane is becoming unrecognizable and it is so because, as Don Draper in the TV series *Mad Men* puts it, "Nostalgia is a twinge in your heart far more powerful than memory alone." The past is not what it used to be, it seems.

Illustratively, we can depict memory as a tapestry, made up of a collection of colored threads, which are the individual memories that together form a picture of the past. The further we move away from it, the less distinct the individual threads become, and the more susceptible each memory is to alteration.

Ever-moving, ever-changing, memories create their own loop, fed by the moving so-called *now* backward to memories that are influenced by later events that change to become another past within a hyperlinked net that drives us from one memory to another in a constantly changing kaleidoscopic turn.

It's not that memories drift back in a harmonious fashion, with older ones neatly being replaced by newer ones. It doesn't work like that. When a memory is accessed, it is altered as it begins its move backward. This newer (altered) memory is fresher in our minds than ones that have not been accessed for a while. It happens every time a memory is brought up in the mind, so that there is a sort of recycling process that constantly brings to the top of the pile old memories that become new, changed memories.

So while it is sometimes assumed that the further one's memory is of an event, the more fixed, and therefore more accurate it would be, this does not seem to be the case. Sadly, perhaps, the notion that 'distance makes the heart grow fonder' is a delusion; this 'fondness' is a distortion, a false memory, of what might have been, and is, rather, what we would have liked it to be. Memories are, according to the great Italian playwright Ugo Betti, "like stones. Time and distance erode them like acid."

Memories cast their spell back and also forward. Why does coffee often not taste as good as it smells? Why, in fact, is expectation often better than the actual event? Once again, we can blame memory and the principle that newer memories are more vivid than older ones. If, as we suggest, *the present* instantly moves back into our memory, the smell of coffee, when encountered, was moved into our memory to be replaced by the taste of coffee, which was itself replaced by the smell of coffee. The newest one always wins. This is backed up scientifically by Aberdeen professor Barry Smith, who shows that swallowing coffee sends a burst of aroma up the back of the nose from inside the mouth, activating a "second sense of smell." It is that second sense of smell that becomes more dominant than the taste itself.

The recollection of a person you haven't seen for a while will lie in your memory as the person you think you remember from then. However 'real' it is, the image will more or less stay like that until you actually see the person, when your memory is 'corrected' and replaces the one you had until that moment.

What happens, though, when the person—let's call him Peter—has changed so much that you do not remember him? With such a big change, the newfound Peter is not matched to anything in your memory and there is no recognition. Had you seen Peter in the intervening period, the gradual changes would have been compared to each other, your memory each time corresponding to the previous one. Memories that are never accessed simply disappear into forgetfulness, perhaps at some point triggered by something in order to be dragged out again, perhaps not.

As a mass of jumbled thoughts mixed in with what we think we remember, our memories are a mess, the results of which are pretty chaotic. For example, we tend to remember unhappy experiences more vividly than happy ones. Linked to that, we brood more obsessively about the dumb things we did, and as we age, we grieve more about all the things we failed to do—or think we did. It can go the other way, of course, and then the dumb things we did are simply wiped out by memories that allow us to live in denial of what happened. "My memory is a liar," mathematician John Horton Conway said. "It's a good liar. It deceives even me."

Hanging on to memories is important for many people. Memories supply a hook onto which one's personality is attached. When we share memories with those close to us, even if we don't agree on all of the details, our memories are reinforced, which in turn reinforces our emotional link with those people. It is not surprising that a loss of a witness to memories is a jolt to one's personality. As Irvin D. Yalom puts it: "To lose a parent or a lifelong friend is often to lose the past: the person who died may be the only other living witness to golden events of long ago."[48] Whether truly golden or not, or whether they seem more golden after the event, those memories have taken a sudden lurch to begin a new path, gradually replacing previous memories.

If there is definitely something in biologist Gerald Edelman's comment, that "every act of memory is to some extent an act of imagination," it should not be a surprise that our self-image is so volatile, hanging as it does on the capriciousness of our moving memories.

Installed memories

All our experiences are potential memories, many of which are dormant until they are triggered by a variety of sensations, that include smells, sounds or advertisements that bring to the surface products that were unconsciously absorbed.

Portraits are a good example of memory etched into our minds by the painter who recreated a memory of what was not necessarily accurate. We 'remember' what Napoleon or George Washington looked like from portraits that reflect the subjective view of artists two hundred years ago, and we access from our memory an image of Jesus as a bearded white gentleman, whose features would not be out of place in Sweden but very much so in the Middle East.

Memories are planted, often even by ourselves. Embellishing the truth to impress friends on social media sites such as Facebook and Tumblr is common. Where it becomes harmful is if it becomes part of who we are and eventually part of our own memory, especially if we can no longer delete whatever we were foolish enough to tap out, all of which remains as a permanent record. Psychologist Dr Richard Sherry, a founding member of the Society for Neuropsychoanalysis, points to studies that verify such modification of memories, to the point of entirely changing peoples' nature over time. The phenomenon is surprisingly common, with a fifth of young people admitting that their online profile, which includes lies about their relationships, promotions at work and holidays, bears little resemblance to reality, resulting in their recollection of past events being distorted by their own fabrications. We manage to fool ourselves as much as we fool others.

There is no doubt that memories play strange tricks on us, sometimes giving rise to, and justifying, strong beliefs. We are prone to find links between memories and actual events, and read significance into them. Often, for example, we are stunned by seeing someone who we had only that moment thought about. The fact that we might have seen that person previously when we hadn't been thinking of him, is not considered; all is covered by the strong feeling of the thought popping in 'just now.' And neither is it taken into account the number of people we think about in the normal course of the day and who do not turn up at those moments. Similarly, possible explanations for the phenomenon of telepathy apply; it does seem uncanny when the phone rings and the person on the line is the very one we were about to ring. Yet, how many times does that happen? More to the point is how many times that does *not* happen. We tend to remember the times it happens, rather than the times when 'telepathy' doesn't happen, which is more often the case.

Spiritualism in general often works on the vicissitudes of memory, and the effects of coincidence and chance. It is easy for experienced mediums to introduce suggested thoughts into the mind, where autosuggestion falls on fallow ground. After all, someone who looks to contact a dead relative or friend is ready for positive results—in fact is primed for them. People who wish to believe are reinforced in their beliefs, tending to remember those details in a séance that were correct, but overlooking statements or events that provide no evidence of paranormal powers. As with telepathy, selective memory is at work here, retaining what did happen and shunting aside that which didn't.

On the other hand, some of our actions are done specifically and consciously to create memories for ourselves. In that way, buying souvenirs is buying memories. Louis Theroux, the British TV interviewer, had that intention when he said, "I did feel as though I'd spent a bit too long in the same house in the same part of London. When you're in your 40s you become more conscious of life being of limited duration and that you need to create memories and go on little adventures from time to time." It's an interesting perspective, expressly creating a past to look back on in the future, thereby consciously changing one's future identity.

Obviously we are treading on some pretty shaky ground here, when we consider the strong connection between memory and identity—not only for individuals, but also for society. Most people consider that memory is an essential part of what makes them who they are, so they are most likely to safeguard and defend their memories—individual, collective and official—in order to hold onto their integrity.

To what extent, though, does what we remember define who we are? Already in the mid 19th century, Nietzsche said that time and Self are bound together, while Freud, who was at first fascinated with the notion that memories give an insight into the person's inner workings, eventually realized that people have strong imaginations that they are convinced are part of their memories. He described memory as a palimpsest, a work of constant reconstruction, and how subsequent experiences can alter both the content and the psychological import of previous memories. This is in line with recent research—for example, evolutionary psychologist Robert Kurzban, who lays out a picture of the human mind as a collection of different parts, or modules, that don't always agree with each other; we can put this down to the constant changing memories that jostle with each other within our minds. The resultant continual reconstruction creates the Self as a work in progress that begins when thoughts form themselves in the so-called *now*.

In the studies on false memories engendered by lying on social media, mentioned earlier, it was found that while seeking support or empathy from our peers is entirely understandable, the dark side of this social conformity is that we can deeply lose ourselves and what it authentically feels to be 'us.' This can, as Dr Sherry says, "undermine the coherence between our real, lived lives and memories." If our concept of Self is simply a bunch of memories bound to time, it is not surprising that our feelings of self-worth are subject to constant change.

It must make us wonder about the notion of the 'true self' and its connection to memory. The American psychotherapist Irvin D. Yalom puts it succinctly:

> Yet, can therapists or historians or biographers reconstruct a life with any degree of accuracy if the reality of even a single hour cannot be captured? ... at such times one longs for an umpire of reality or some official sharped-imaged snapshot of the hour. How disquieting to realize that reality is illusion...[49]

The notion that there is a 'true self' is, then, dubious. How could one even attempt to find it? With psychology no longer so certain that one's true self is a repository for one's devils and past history, it would be fruitless, perhaps even harmful, to examine memories as keys to one's 'Self.' After all, if a memory is an illusion similar to, say, a mirage, how can we base what we are according to something that may or may not exist?

Yet, memory is often dramatic and powerful. In Rachel Joyce's *The Unlikely Pilgrimage of Harold Fry*, there is a wonderful description of Harold Fry's state of mind as he takes his monumental walk through England, when we are told that "He no longer saw distance in terms of miles. He measured it with his remembering."

It's an interesting point, and the more we think about it, the more it validates itself. We can, surely, measure things only according to what has already happened. What is there to hold onto, other than a past—a past that is inexorably moving back into memory?

The notion, very strongly felt, that one is able to sort out real memories from one's swirling cauldron of thoughts, real or imagined, needs to be seriously questioned. In fact, the very idea that there is such a thing as a 'real memory' is not on. This has been apparent for some time. French historian Pierre Nora, who worked extensively on French identity and memory, summed up his view in these words: "Memory is in a permanent state of evolution, open to the dialect of recollection and amnesia, unaware of its successive deformations, vulnerable to all the ways in which it is used and manipulated, susceptible to long periods of latency and of revitalization."[50] In other words, beware of relying on memory.

Also, the idea, popularly believed, that memory doesn't change—but merely our perceptions of what happened—leans on the assumption that there is such a thing as a 'real memory.' It is the notion of memory being a tape recorder or CD that plays back perfectly every time you press 'Play.' The ultimate question is whether a 'real memory' is of an event that actually took place; it could easily be a really powerful image in one's mind of something that never actually happened. We will look at this when we discuss false memories.

Unlike that which is typed into a computer, which even if deleted could be reconstructed, if human memories are deleted they could never be reconstructed, since there would be no indication as to how close they are to what had been there previously.

The whole issue of memory is a minefield. It becomes particularly hazardous when traversing it with a mine detector that may not always work.

Not knowing how much of a memory is genuine or even if it is a memory at all can shake up one's sense of reality. Paul McCartney relates that he composed the song 'Yesterday' in a dream and when he awoke, he immediately played it on the piano so that he wouldn't forget it. After that, he began to worry that he had taken it from somewhere, doubting his dream and his memory of the dream. To settle his doubt, he sent the composition to various bands, all of whom said that it was, indeed, original.

The question of plagiarism is a subject fraught with murky

unknowns. It is a mark of shame in schools and universities, in academic papers, journalism, music and a host of activities where originality is an important part of output. Unconscious plagiarism, a sort of mental kleptomania, is a particularly pernicious occurrence that plagues many a composer of music or written text. How can one be aware of a memory that one is not aware of? How does one compose music without the possibility that a remembered tune didn't pop into one's mind, or produce a text without an idea that one had read sometime in the past?

It happened to George Harrison, whose song 'My Sweet Lord' was released on January 15, 1971 and hit the charts eight days later as his first solo single. A few weeks later Bright Tunes Music Corp filed a suit against him, claiming his song was the same as 'He's So Fine' recorded by the Chiffons in 1962 and later moved under the Bright Tunes label. George Harrison was found guilty of "subconscious plagiarism' and a judgment was filed against him in the amount of $587,000, which was paid in full. In other well-known cases, such as Michael Bolton vs. the Isley Brothers, Johnny Cash vs. Gordon and John Fogerty vs. Creedence Clearwater Revival, unconscious plagiarism possibly took place, as well.

For writers, whose careers rely on reading to bring out ideas, unconscious plagiarism is a professional hazard. How can they know whether an original thought is not, in fact, a remembered segment brought to the fore when 'pen hits paper'? There are a host of writers who discovered to their chagrin that part—sometimes large chunks—of their work had been lifted unconsciously from others: Maureen Dowd, Chris Anderson and Elizabeth Hasselbeck were guilty, as were, earlier, Nietzsche, Lord Byron, Robert Louis Stevenson and a host of others. Umberto Eco was mortified to discover that he had copied the whole opening passage from *Name of the Rose* from an antique book he had bought in his youth, parts of which had lodged in his memory.

That brings us naturally to déjà vu, the uncanny feeling that you remember a situation from before you encounter it for the first time—the mind-boggling perception of having remembered the future. Leaving aside the entertaining explanation of 'a previous life' being remembered, the phenomenon of déjà vu fits in nicely within our discussion of memory, time and *now*. Our memories are of what occurred (or, perhaps more correctly, what we believe occurred). As we have seen, when something happens, it actually does so in an instantly backward-moving *now* into our memory. When we recognize something, we recall it from our memory and match it to what we have come across. So what our hard-working mind is doing constantly is retrieving memories or forming new ones. With the continuous array of memories that either do or do not match up with what we come across, it should not be surprising that now and again a false memory is 'retrieved' from what it had only that instant experienced, giving the distinct feeling that it had already occurred, further back in time. So in the infinitesimal time available, instead of forming a new memory, our mind recognizes what it falsely considers to be an existing memory. A glitch in the circuit, to put it plainly. The result is that the déjà vu memory becomes a normal memory, joining the other memories as they slip again into the string of memories.

In all of the jumble of memories are those that people seem to remember forever: the memory or memories that have had such a traumatic effect that they are emphasized as though enclosed within a series of exclamation marks, the slow-motion series of images of an event that took only a few seconds in real time. Journalist John Hockenberry relates a gripping story of the car crash that changed his life, after which he was never to use his legs again, and the excruciating detail-by-detail memory of that event.[51]

In the previous chapter, we mentioned subjective time

in which events are perceived as slowing down. It was the neuroscientist David Eagleman who showed that the passage of time can be subjective, by demonstrating that time itself does not slow down even when people were convinced it does. Since time seems to slow down in times of stress, fear or danger, Eagleman devised a test that set out to show what happens to time in those conditions, by taking volunteers to a 150-foot tower in Houston, Texas, where they were dropped in free fall at about 70 mph and caught in a net below. The volunteers were given numbers to read that were flashed at them just too quickly for them to be readable. The point was that if time slowed down for the volunteers as they dropped, they would be able to read the numbers. But they weren't able to. Time did not slow time, even though the volunteers were sure that it had.[52]

What causes this phenomenon of perceiving time as slowing down? Eagleman claims it is due to memory, where traumatic events are retained in the mind in greater detail than other events, so that when replayed, the remembered details are stretched as if over a long time. A trick of memory, in other words.

It makes sense, yet speculative in that there is no way of testing whether (or why) memories are retained according to effect or importance. After all, we don't know which important events we have forgotten. Not to mention that often events are so traumatic that they are completely erased from memory, as when people say that they recall nothing of an accident and remember only suddenly being in hospital.

There is another explanation as to why traumatic events seem to take a long time: distorted memory. A false memory, in other words. Composed of details that the person might have heard, or pieced together from what is seen immediately afterward, the result is an imagined memory of details that *could* have taken place. All details sound reasonable and the person will be absolutely certain of his vivid memory.

It's a harsh theory, since it undermines those important

memories of the traumatized victim. Yet, why would some memories be more reliable than others? And how could we possibly know?

False memories

The question as to whether or not forgotten memories can be brought back—and *if* they can, how reliable they are—touches directly on two important fields: psychology (particularly psychotherapy) and the justice system. They are, of course, connected.

Courts of law, especially criminal law trials, need to deal with memory in almost every case, where probing witnesses in order to jog buried memory is an important element. Identification in a police line-up is routine before a case comes to court and is an important peg in securing convictions.

The high number of wrongful convictions based on testimony relying on memory that were later found to have been inaccurate, obviously sounded a number of red alarms. The problem was not that of deliberately lying, but of memory that had been altered through circumstances of which the witness was not consciously aware. Recognized as a condition that needs attention, False Memory Syndrome has become an important issue, bringing courts to rely on forensic evidence more readily than on eyewitness testimonies, not to mention admissions by defendants to crimes they had not committed. It had also become apparent that false memories were often triggered by the way witnesses were questioned by counsel. Neuroscientist Steve Ramirez, who has done important work on memory manipulation, has stated that "Memory is not a carbon copy, but rather a reconstruction, of the world we've experienced."

Detailing hundreds of cases of wrong convictions, the Innocence Project was set up in the USA to exonerate wrongfully convicted individuals, while cognitive psychologist Elizabeth Loftus, already in the 1970s published a series of influential studies about the problems associated with eyewitness testimony. She cast a spotlight on false memories, while challenging the accepted belief that traumatic memories can be buried and recovered, and making the startling assertion that her search into the scientific literature did not find any convincing evidence for the claim.

Memories can be planted, and can be done easily, as shown by Loftus, who has spent her career deliberately planting false memories into the minds of the people who volunteered to take part in her studies. "We've been planting what we call *rich false memory* for decades now," Loftus said. "We and others can get people to believe these very detailed events happened to them—that you were lost in a shopping mall and frightened and crying, that you nearly drowned and had to be rescued by a lifeguard, that you were attacked by an animal." In an aptly titled article 'A picture is worth a thousand lies: Using false photographs to create false childhood memories,'[53] a research study is detailed in which half of the participants have partial or complete false memories, sometimes even 'remembering' additional details from the event—an event that they never experienced—after seeing a Photoshopped image of themselves as children, standing in a basket of a hot-air balloon. "It's pretty normal," said Deryn Strange, an associate psychology professor at John Jay College of Criminal Justice, New York. "That's the hard thing to get our heads around. It's frightening how easy it is to build in a false memory." Or as Elizabeth Loftus puts it: "Memory, like liberty, is a fragile thing."

It's not all necessarily grim, though, and it's not always as sinister as in the movie *Total Recall,* in which Arnold Schwarzenegger attempts to have false memories about a holiday in Mars implanted into his brain in order to disentangle memory and reality. Outside of the movies, the malleability of memory has potential advantages, so that tampering with it—planting false memories—is sometimes used to the good, such as causing people to eat more healthily, or treating alcoholism and posttraumatic stress disorder after rape or war. Whether it is called 'planting false memories' or the easier-on-the-conscience 'rewriting memories,' the act is appealing. Edna Foa, director of the Center for the Treatment and Study of Anxiety, at the University of Pennsylvania Medical School, and one of the leading experts on the psychopathology of anxiety orders, states that rewriting memories, rather than destroying them, is a worthy enterprise. "Nobody is suggesting," she says in an interview, "that we rewrite the memory of someone who had a bad date or a fight with their mother."[54] But it could be done.

Appropriately, human memory has been compared to something more akin to a Wikipedia page, in that you're able to make edits—and so can other people—to your own memories. All this in addition to the edits that are going on automatically, as we have seen, as change that occurs through time

For reasons that are as mysterious as the memories themselves, people have fabricated rich and detailed memories of events, such as having been abducted and impregnated by aliens (claimed by an estimated 3.7 million Americans, and serious enough to cause nightmares and other problems), or that they were Napoleon in an earlier life. They are absolutely sure that their memories are genuine. And there is no reason why they should think otherwise, since the brain does not physically distinguish between what could be *real* and what could be *false* memories. This should not surprise us, since, as we have seen, the notion that there is such a thing as an objectively real memory is highly suspect.

Memory can be dramatic and sometimes very much so. Memory researcher Doug Bermingham works with war veterans who have misremembered some incredibly dramatic events, wrongly blaming themselves for what happened at events in which they were present.

The rub is that we don't know how greatly memories are

being modified, although stress or pressure could cause drastic change. "Misinformation," Loftus wrote,[55] "has the potential for invading our memories when we talk to other people, when we are suggestively interrogated or when we read or view media coverage about some event that we may have experienced ourselves."

The idea that misinformation can invade our memories is all too common. Lawrence Patihis, a psychologist researching memory at the University of California, Irvine, mentions that in one of his studies, about 20 percent of participants generated false memories of an event after something false had been suggested. For example, if Patihis mentioned video footage of Flight 93 crashing in Pennsylvania on September 11, 2001, one in five participants said they remembered seeing it. No such footage exists. The power of suggestion is great indeed.[56]

Suggestion builds into memory over time. In experiments set up with séances where spirits were expected to lift the table, two weeks later many of the participants remembered the table rising, even though nothing had happened at the time. Once implanted, it is a memory and will remain so whenever it is accessed.

Implanting false memories sounds like a dream come true for autocratic regimes, as it is for cults, who manipulate people's personality and sense of Self. A classic example is Scientology, whose system of Dianetics aims to clear a person's mind of all painful and harmful memories in a process called 'auditing.' The 'eternal happiness' that follows is the result of a mind that is completely open to the cult's follow-up manipulation and enslavement.

Making sense of memory within the individual is a nightmare. While it is clear when memories are planted from outside malevolently, or in a laboratory as an experiment or for benign reasons, we get into a foggy area when it comes to repressed memories. Is there any way we can answer the question we asked previously: Is it possible to bring back forgotten memories? With all that we know about reliability of memory and the influences that are active after the possible event, every answer necessarily spawns more questions; here's one, for a start: How would we know—how could we possibly know—that a memory we don't know about was forgotten? If sometimes it's tempting to play with the idea that the best memories are those that we have forgotten, we would be shooting wistfully at the nonexistent. A memory that we have forgotten is not a memory.

Many of these questions came to the fore in the 1980s and 90s during the spate of announcements by celebrities of remembered child abuse within the family. Although these have abated somewhat since and are no longer so newsworthy, the whole issue of repressed memories is being struggled over. Considerable energy and anger is expended, with both sides of the reality divide bringing evidence to make their case. Particularly vociferous are those who claim that denial of repressed memory allows the perpetrators of abuse to remain unpunished, and worse: making out that the victims are deceitful.

Yet, the stream of evidence from experts in the field of memory continues to pound the message that the very notion of repressed memories is suspect. Books with titles such as *Remembering our Childhood: How our Memory Betrays Us*[57] and *Remembering Trauma*[58] help to keep the issue alive, aided by Meredith Maran's memoir, *My Lie: A True Story of False Memory*,[59] which was particularly poignant. Ms. Maran had accused her father of molesting her, based on a combination of 'symptoms' like depression and guilt, and disturbing incest dreams, provoking an estrangement that kept her children from spending time with their grandfather for eight years. It was only afterward that she recanted and told the world that it had never happened.

The thing about memory is that for the person who has it, it is absolutely real. When a memory comes to the fore, there is no attempt at conning anyone. And this is the tragedy, not least because any attempt at telling someone that what they remember did not happen may have no effect other than angering the person. It will possibly not get them to think that their memory is not real, even if you bring contradictory evidence.

There are widely reported cases of public figures whose memories turned out to be more imagination than recollection. Ronald Reagan, in the 1980 US presidential campaign, often related a touching story of a World War II bomber pilot who, when ordering his crew to bail out of his stricken plane seriously damaged by enemy fire, saw that a young gunner was wounded so seriously that he was unable to evacuate the plane. Reagan could barely hold back his tears as he repeated the pilot's heroic response: "Never mind. We'll ride it down together." As touching as it was, the press soon realized that this story was an almost exact duplicate of a scene in the 1944 movie *A Wing and a Prayer*. Reagan had apparently retained the facts but forgotten their source.

Reagan was not simulating emotion when he recounted this story—his story, his reality, as he believed it to be, was part of his memory. Except that it wasn't. Had he taken a lie detector test, he would have passed it, since there was no conscious falsehood that would have shown up.

Statesmen seem particularly prone to memory imagination, but that might seem so simply because they are in the news. Hillary Clinton mentioned a memory she had of a trip to Bosnia in 2008: "I remember landing under sniper fire. There was supposed to be some kind of a greeting ceremony at the airport, but instead we just ran with our heads down to get into the vehicles to get to our base." When it was pointed out to her that the actual trip, with her daughter, Chelsea, in 1996, was far different to what she remembered, and that they were greeted with hugs and gifts, and no sniper fire, she said: "I made a mistake. I had a different memory… I'm human."

A similar memory lapse very nearly cost NBC anchor Brian

Williams his job when, in 2015, he was discovered to have made a false claim about being in a helicopter in Iraq that was shot at. "We all misremember details and experiences from our past," Benjamin Radford commented in an article, "and most of the time we aren't aware of it because we don't verify our recollections against the memories of others who were with us at the time, or compare them against some objective documentary evidence of what happened such as a videotape or news report."[60] More interesting than the comment about all of our memories being suspect in some way, is the assumption that evidence can be objective. It can't be, of course. That there is a 'real,' 'true' memory out there is, as we have seen, essentially an impossibility.

Memories can be altered by even the most innocent and otherwise beneficial of practices, such as mindfulness meditation. Its alleviative effect on stress reduction and a host of other benefits is, as recent research has found, countermanded by the downside effect of making us more likely to 'recall' something that never actually happened. By having their minds wander freely to embrace judgment-free acceptance, "meditators can have greater difficulty differentiating internal and external sources of information," concludes a research team led by University of California—San Diego psychologist Brent Wilson.[61] "Their reality-monitoring accuracy may be impaired, increasing their susceptibility to false memories." What happens in mindfulness meditation is that the mind ceases to hold onto clues from memories, and is a sieve (literally 'mindless') that is susceptible to suggestions from its surroundings. One's memory, in fact, has been wiped clean to a large extent, a situation that leaves it open to accepting memories that are false.

While it is understandable—normal, even—for memories to be changed naturally, the situation becomes extremely worrying when we consider the people who have created a memory of something far removed from anything that ever happened.

In the 1990s, a Swedish man calling himself Thomas

Quick (whose real name was Sture Bergwall) confessed to 30 gruesome murders, spanning two decades. In therapy sessions at the psychiatric hospital, he said that he had killed, maimed, raped and even eaten his victims, the youngest of whom was a 9-year-old girl. He blamed it on the horrific sexual abuses by his parents. While the confessions of a real-life Hannibal Lecter were stunning Sweden, they turned out to be a figment of his imagination. They were all false memories, including his abused childhood. Until this was uncovered, he spent 22 years incarcerated in hospital until he was released in 2014. "It's the judicial scandal of a century," says Dan Josefsson, the author of a book on the case, "and it all has to do with a therapeutic idea that was in fashion in the nineties." Thomas Quick unknowingly fooled everyone with one confession after another in order that he could stay in hospital (which does seem to be the best place for him, even if he is not a murderer).

The writer Will Storr[62] relates the case of Carol Felstead, who died alone after years of being convinced that she had suffered Satanic child abuse at the hands of her parents. Carol—by now Carol Myers—had left a long note that stated: "My parents were abusive in every way imaginable—sexually, physically and emotionally. At three years of age, my mother smothered my sister. She sat me on top of her body and set the house on fire." None of this was correct, as discovered after lengthy investigations in which it was shown that her sister died before Carol was born and that the house fire also predated Carol's birth.

The whole notion of therapy that supposedly recovers memory—RMT, recovered-memory therapy—has turned out to be hugely problematic, with a charity organization, the British False Memory Society, listing cases where people have been falsely accused of rape or abuse.

Some of the cases are quite hair-raising. A famous and bizarre example is that of 'Sybil' and the more than dozen personalities she claimed lurked inside her, in addition to the tales of abuse suffered under her mother. After a book and TV movie came out about this in 1973, it was exposed in 2011 as a complete fraud on the part of the psychiatrist, who treated her with drugs, coupled with an unhealthy relationship, which brought out memories that were false.[63] The controversy about 'Sybil' (Shirley Mason, her real name) concerning which version is true, partly true or completely untrue is in itself interesting, since there is no way of getting anywhere near what really happened. At one stage, though, Mason admitted to her psychiatrist that she had been lying. Not accepting her admission, her psychiatrist believed instead that Mason did not wish to continue with therapy!

A case that brought huge and irreparable monetary and familial damage was that of the Gary Ramona family, where the daughter, Holly, revealed in therapy that she had been horribly abused and repeatedly raped by her father, and had been made to have sex with the family dog, etc. In 1994, Gary sued the therapist and won.

Holly had entered psychotherapy five years earlier with serious symptoms of depression and bulimia. Over time, she spoke about her father's molestation, that became evermore graphic, culminating in the scene of a rape. Not knowing whether to trust her own mind, the tormented 19-year old persuaded her therapist and a psychiatrist to give her sodium amytal, that triggered increasingly perverse 'flashbacks' from age 5 to 16. Rejecting her father's protestations of innocence, Holly refused further contact, while her mother took her two sisters and sued for divorce, and her father, a wealthy wine executive, losing his half-million-dollar job.

The wide gap between memory and history (that which is thought to have happened) is shockingly apparent. Behind all of this, though, is the thorny issue of 'truth' and what it could be. As interesting as the details of Holly's case are, and as much as we would like to know 'what really happened' (after all, that is what it's all about), it is clear that all sides were sincere: they were telling the truth and honestly remembering things that never happened. It's not, then, a matter of 'lying,' which is what Holly's mother and the therapists wanted to get to the bottom of. Holly's memories could have been the result of a number of traumas that have to do with growing up, and not uncommon with sufferers of bulimia.

This, though, is not the place to go into the merits of the case or to list others, many of which have been widely discussed in the media, including those where recovered memory has been discredited, as in the case of Cardinal Joseph Bernadin, charged with sexual abuse by a former seminarian and who subsequently withdrew his allegation. Obviously, not all 'repressed memories' are necessarily false. But that is not the point. Or at least it is only part of the point. What we are looking at is the nature of memory and what our memories reflect.

It was Sigmund Feund who brought to the fore in the 19th century the notion of repressed memories and the supposition that they are harmful. Psychological therapy is based on the premise that facing one's memories is beneficial. Yet that might not always be the case. It makes no difference if those memories are implanted ('false memories'); they are 'real' and vivid, as strongly as other memories, no matter the source. Take the breakup of Yugoslavia in 1992, and the subsequent deadly fighting between Serbs and Croats, neighbors who had been living together in one country before the Serbs threw hundreds of thousands of Muslims out of their homes, whipped up by nationalist leaders who uncovered memories of a dubious past. If ever there was a case of cynical manipulation of so-called memories, here it was, resulting in horrendously deadly consequences.

The point here is not just the uncovering of memories, but also of memories that were based on events that did not occur (or may not have occurred). The Serbs had been convinced by a cynical leader that these secular Bosnian Muslims, so recently part of the same country called Yugoslavia, were a reincarnation of the Turks of old, latter-day Ottomans determined to take over Christian Europe. And so, back at each other's throats, Bosnians, Serbs and Croats re-enacted, in the 1990s, the horrors of the civil war of the 1940s, whose mention had become taboo—and rightly so, it turned out—under the clamp of Tito's postwar Communist dictatorship, and that had again been brought to the fore by cynical and manipulative leaders.

All memories are felt to be real, whether they are recalled 'repressed memories' or 'false memories.' How and what triggers them is immaterial.

Often, our past is thrown back at us when it had been dead and buried, and we wished it had stayed so. What was once a harmless photo with a then-partner, or a sexually provocative private photograph, can have the embarrassing misfortune of turning up years later, out of context and therefore changed. Such was the incident at Glastonbury in 2015, when a festival-goer stood in the crowd during Kanye West's show, waving a homemade banner depicting Kim Kardashian giving oral sex to her ex-boyfriend Ray-J, taken from Kardashian's sex tape, released without her consent in 2007.

However long ago it was, however much you've changed, there aren't any clean slates that stay that way forever. With the ubiquitous smartphone, it doesn't require time or knowhow to snap a photo at any time. And, of course, there is sexting—the explicitly sexual messages and photos, sent mostly through smartphones—so widespread as to get into the news only when some serious harm has been caused. A memory best forgotten, it isn't allowed to be when the objective in bringing it out is to harm someone.

Despite memories being a natural and important part of being human, releasing all of them is not always healthy, especially as they are likely to be subjectively altered in any case. Releasing devils is not good for anyone.

Holding onto memories

So where are we? Stuck within the imponderables of memory, it seems. How could we possibly know what our 'real' memories are? Some—or at least those who have thought about it—have tried to find out. One is French poet Jacques Roubaud, who spent much time writing out his memories, filling volumes that were published, and never looking back over what he had written. The object, he said, was to discover his *own* memory by simply describing the memories that came to him. By doing so, he maintained that he could 'destroy' any false memories before they had a chance of occurring.[64] The problem here is that he was defeating his own purpose, since by not looking at what he had written down, he would not be able to refer to what those supposedly real memories were. He was definitely in a bind, since if he ever did decide to look at what he had written, the very action of doing so would likely affect those memories, again defeating his original purpose. In other words, marking out your memories is not a way of pinning them down as genuine; you won't know anything about them if you subsequently don't refer to them, and you will change them if you do. Damned if you do and damned if you don't.

Basically, though, that is what diary writing is all about: inscribing events or thoughts in order to preserve them, so that at a later date we can refer to them as our memories. As a record of our memories that's only partly effective, of course. There is a very good chance that we will have different feelings about the things we had written about at the time. In many ways, we won't be the same people, having had experiences that morphed into memories, so giving a different nuanced meaning to our written records, without even realizing it.

One way of getting past this could be to write down something important that has just happened "while the memory is fresh," as advised by Craig Good in his Skeptoid blog. He also advises us to "take a photograph or video. Stay humble. When you catch yourself saying, 'I know what I saw!' step back and realize that maybe you don't. See if you can corroborate it with other witnesses or records. And, as always, be skeptical—especially about things you are just sure you remember."[65] That seems eminently reasonable.

Yet it isn't. Asking us to question our memories is barren. The notion that we can verify our memories by making notes when things happen that we can check afterward is a delusion. We cannot capture a moment, so that even as we write our notes, we are coloring what happened through the so-called *now* that is moving into the past and is already a memory.

Similarly problematic is the advice to journalists, in an article in *Slate*,[66] to "check Google to verify any events before publishing." But as verification of what actually happened at past events, Google is no less subjective than other sources and certainly not more reliable than the other advice to "use your personal archives. Go back to data stored in your own smartphone, computer, and cloud accounts to find corroboration for your memories. The web of technology that envelops us provides valuable, time-stamped documentation of our thoughts and actions." But all of these rely on the vicissitudes of memory. Time-stamping tells us only *when* something happened, not what *did* happen; it certainly does not stop changes in memory from then. Time and change do not stop.

So perhaps a solution would be to write a live diary, a journal describing what is actually happening at the time it is happening? At first glance it does seem a reasonable way of holding onto memories. But think of what that would entail: it would take the diary writer longer than a day to describe the day itself. And not only would they be struggling to keep up with events and thoughts, since there would hardly be time to do anything else, they would, in fact, be doing nothing else other than writing, hence would be writing about themselves writing. Extreme accuracy of what was happening would turn into a parody of itself, resulting in the opposite of what was intended. And videoing instead of writing would not make the whole effort easier; it would mean someone videoing you while you were unable to, eventually reaching the same situation as physically writing it all down. The whole venture is absurd. Making a live diary (whether written or recorded) manageable would necessarily entail leaving much out. The paradox is that leaving out details makes memories approximate and therefore less accurate. And that, surely, is a characteristic of preserving memory: a process of leaving out facts that would clutter up the whole but with a resultant less accurate picture.

We are left with the paradox that the more accurate something is, the more useless it is in practice. If we think of a diary as being a map of our memories, we find interesting relevancies. When we need a really accurate map for, say, hiking or cycling, we would go for one that is to the scale of one-centimeter-to-the kilometer. That would be the most accurate map one could acquire. As accurate as such a theoretical map is, the map would still have to leave things out. To keep maps manageable, and to avoid cluttering and overlap of detail, cartographers can include only a small proportion of the information on a landscape. In theory, the map that would most accurately represent the territory would include everything and would be, as Jorge Luis Borges pointed out in a short story, a map on a scale of one-to-one. When Borges's ambitious fictional mapmakers tried this, they ended up with a sheet as large as their empire and the whole enterprise collapsed.[67]

The paradox manifests itself in relation to using GPS versus a map, and whether different types of accuracy relate to usefulness. Whereas using GPS gets us from one place to another and tells us where we are, it tells us nothing about our location in relation to our surroundings and what is nearby. A map, on the other hand, while enabling us to plan getting from one location to another, will also allow us to see what is around and where we are in relation to other places. In a similar vein is the analog watch face versus digital timekeeping, where the digital clock tells us what time it is but does not show where we are in relation to the time before and after as indicated within the time markings on the clock face.

All this seems to indicate that it is often more useful to know ABOUT something—a representation that shows relationships—rather than attempting to know what that something IS. In other words, we know more about something when it is placed in context, relative to other things. And so it is with our memories: we can only know *about* them, not what supposedly 'really happened.'

Unlike a map, though, which we can check to see if it is accurate by going to the territory it represents, we can never check the accuracy of our memories. As we have seen, we would be on slippery ground, indeed, if we insisted that memories are of what actually happened, however much we feel it to be so. Unlike visual art, for example, where sometimes 'less is more,' so that leaving out detail gives the observer room to fill in the gaps with their own imagination, with memory we would obviously not expect imagination to be part of it, if we insist on it being a reflection of what actually happened. Yet, with memory being so flimsy as a recollection of possible past reality, imagination plays an important role. "It's not surprising," says Daniel Schacter, a psychologist at Harvard University, "that we confuse memories and imagination, considering that they share so many processes." How, then, do we know where the line is between memory and imagination?

It is normal to forget and given the tenuous nature of memories, it is not surprising and may even be useful as a psychological tool for survival. What about people who claim they never forget anything? These are such rare cases, one being Jill Price, a 34-year-old housewife, who claims that she can recall key events on any date back to when she was about twelve, as well as what she herself had done each day. She was not exaggerating when she said that "I run my entire life through my head every day and it drives me crazy!", for laboratory tests showed that she was right. She really didn't forget anything; she was able to state what had happened historically on any particular date, which was checked against a copy of *20th Century Day by Day*, a book that lists important events by date.[68]

The fact that she, and others with this condition called hyperthymesia, remembers recorded events is an interesting and rare phenomenon. The fact that she failed badly at other tasks, such as remembering numbers or words, should make us wonder whether or not the memories she did have would be less subjective than anyone else's. We would surely not claim that her memories of an event would be objectively correct, her mind a metronome of real memories.

For most of us, who do not remember everything, photos are always useful to bring us back to the past. Or so it seems. It's tempting to compare memory, a time-sensitive dive into largely murky territory, with photos, those seemingly solid, ageless reminders of what had been. Feted photographer Hiroshi Sugimoto called photography "the fossilisation of time." The comparison is not apt, though: a photo can lie as much as a thousand words. Photography doesn't freeze time. Time is change, and nothing stays as it was. Look at a photo of people who we know have subsequently died. We don't—we can't—look at that photo objectively, our minds clear of what happened to those people subsequently. Impossible as it is to rid our minds of what we know happened to them (or to anyone on a photo), it's as if, when looking at the photo, we are peering into the future, foreshadowing events to come. Our perception is colored, with the past and the future turned upside down. It questions and shakes the truism that we remember the past but we don't remember the future; by looking into that photo, our memory has been altered and so the static images have themselves been altered: it's as if we are remembering what is still to come. Remembering the future!

Photos turn out to be disconcertingly spongy. Photos depend upon who is looking at them, changing according to the beholder. A photo of people we knew, spent time with or with whom we had a relationship is seen completely differently by someone who did not know them. It's as if they are different people. For us, we are looking at a memory, for others it's looking at strangers who are not part of their memory. That one photo can be considered as two photos, both real but in different ways. How often it occurs that we show, or are shown, a photo of someone and say that 'it doesn't really capture the person.' Of course it doesn't; it never can if the image is of a memory that is constantly changing.

The very notion of a frozen instant in time is rather suspect. Looking at a photo of someone and comparing it with that same person (or a photo of that person) a few years further along, it will be alarmingly clear that every facial line and hint of a wrinkle had subsequently been deepened and exaggeratedly grooved over time, so giving us the uncanny sense of being able to look into the future. This can be felt again through another exercise: by looking at a recent photo of someone else and imagining the lines in their face exaggerated, you will get a glimpse into the future as to how that person probably will appear. This is what computer-aging programs do, remarkably successfully. As a sense of the past and the future fuzzing their autonomies, we can appreciate how memories are susceptible to change and so are far from being as clear-cut as we often feel they are.

There is another problem with the instant captured by the click of a camera. Because it captures a still moment, never normally seen divorced from the continuously moving body, it captures a unique expression on a person's face. The expression is in that sense as 'false' as it is 'true,' for it is supposedly a frozen point in time. Yet, there are no points in time, which is a continuum. Is that a 'now' instant, perhaps? Obviously not, since it has passed. So could we say it was a *past* 'now' instant? Here again, obviously not, since that would be a memory, affected by what you know came later and no more frozen than any 'present moment.' There can be no stop to time, and no way that it can be stopped for us to look at.

It goes without saying that every photo is a reflection of the past, whether of people or not, and then becomes part of our own personal memory. Seeing photos of the destroyed buildings in the aftermath of war or of a terrorist attack brings out our imaginations of what had been, just as strongly as seeing images of bloodied corpses. Photos that depict domestic objects whose meaning has been altered in the aftermath of a calamity, jolt the comfort of memory; the mounds of discarded suitcases in German concentration camps of WWII, is an image that turns domestic memory upside down. An image of a ruined kitchen, like the one by the Ukrainian photographer Sergei Ilnitsky, taken in August 2014 in Donetsk, a major city in the eastern part of Ukraine, did as much to bring to the world the facts about Russian involvement in Ukraine in 2014. It brought out a memory from our own file of memories, of a kitchen but twisted and unnaturally empty, like a bad dream.

The fact that an object connecting us to our personal everyday existence can often evoke stronger feelings than photos of horrific events themselves, is a manifestation of backward-moving memories, the more recent ones overlaying earlier ones, beginning from the instantly moving back-to-memory *now*.

Knowledge

Knowledge is the result of what we know; the result, in fact, of information that we have picked up along the way. Knowledge is not—cannot be—static; it is a collection of accepted facts, constantly being added to and changed, along our moving time into our memory. Knowledge is, to all intents and purposes, memory.

Popeye can help here—he and the accepted fact that spinach is extremely rich in iron, more so than other green vegetables. Unfortunately, that long-accepted information turns out to be fluid to the extreme, since it is based on an error made by German chemist Erich von Wolf in 1870 when analyzing the iron content of green vegetables. What difference a dot makes! It took one tiny slip when transcribing data from his notebook to misplace a decimal point, resulting in spinach being reported as containing a tremendous amount of iron—35 milligrams per serving, not 3.5 milligrams (the true measured value)![69] And so the legend of spinach's nutritional power took hold, becoming evermore popular and influential over the years, in spite of the fact that the error was in the meantime corrected, in 1937. The error did not stop studio executives choosing spinach as the source of Popeye's vaunted strength, that continued well into the 1990s. It would have been a shame to shake the accepted knowledge of the nutritious value of spinach, since it was a favorite food of American children, who boosted spinach consumption by 33 percent during the Great Depression. Knowledge is knowledge, after all. A full-color Popeye statue was erected in Crystal City, Texas, in 1937 (the same year the decimal error was corrected), to celebrate the boost to the region's spinach-growing industry.

This is an example of general knowledge that was inaccurate, is it not? No, not really.

It is, surely, more accurate to say that it is an error today. At the time—from 1870 to 1937—it was accepted knowledge upon which commercial and domestic decisions were made, causing people to eat more spinach. Of course, at any time during that period, if someone had checked, the error would have been discovered and that particular piece of knowledge would have changed. But it wasn't. Not until 1937.

The question is whether it was true at any time in the past.

And if it was true, was it a fact? In short, do facts change over time, or is a fact deemed to be so eternally and objectively?

Intuitively, we like to define a fact as a reality that cannot be logically disputed or rejected, and thus is objectively true and always will be. So was it true that spinach had enormous nutritional value? Obviously, it isn't true according to what we know today, even though it was true at the time according to the facts known then. Assuming that facts are true, was it a fact then, but not a fact today? That gets us into immediate potential difficulties regarding our knowledge today: since facts may turn out to be different tomorrow, we cannot in all honesty be so sure as to call any knowledge a fact. It happened with the nutritional 'fact' of spinach and it happens in science all the time. After all, it was known to be true for some 1400 years that the Earth was a fixed, immovable mass, located at the center of the universe, and that all celestial bodies, including the sun and the fixed stars, revolved around it. New evidence comes to the fore and past findings are discovered to have been misguided or erroneous. Every day we discover that what we thought to be true, is not true at all.

In the light of what we have pondered regarding the nonexistence of *the present* and the backward movement-into-memory nature of *now*, we need to look at 'facts' and 'truth' in a more generous way. It should come as no surprise that facts and truth are not static. Today's 'facts' and today's 'truths' are no more fixed than yesterday's, than last year's, than last century's or than any time in the past. In a nutshell: "There are no facts, only interpretations," to quote Nietzsche.

So the big question is: Was it true that spinach contained a huge amount of iron (35 milligrams per serving)? Or that the Earth was a flat sphere, a belief seemingly prevalent among ordinary people for thousands of years? Or that not so long ago, the Earth was considered to be 6000 years old and that heaven floated somewhere above our heads?

The answer would have to be that, yes, it was true—at that time. It was a known fact and so it was true, and justifiably so, since it was taking into account the information known then. Knowing what we know today does not change retrospectively what was not known then. Knowledge is generated by people, so it evolves over time. Our knowledge today does not affect what people knew in the past. Something that is unknown cannot be true, since there is no one to know it. And the fact that our knowledge is different today doesn't mean it will not change in the future.

It goes without saying that we can only know what we already know. Now although that sounds circular, the point, as obvious as it sounds, is that we only know something that has already happened. In other words, we know what we *knew*, just as children before 1937 knew that spinach, more than any other vegetable, made them strong.

Knowledge is based on memory, as we pointed out earlier. Knowledge, indeed, *is* memory and as such it is not surprising that it is a moving, constantly changing phenomenon that can only be considered as temporary.

The Shakespearean claim that a rose by another name would smell just as sweet makes sense only to people who have an inkling of what a rose is and what it smells like. We can recognize a rose only because it is in our memory. Memory supplies us with what we call the 'familiar.' That which is not familiar is not in our memory, but becomes familiar as soon as it moves to our memory.

So if, as mentioned, we know what we *knew*, the search for new knowledge is, in fact, an attempt to add new information to our memory. It's a search toward what we don't yet know so that we can add it to the past. The ancient saying, 'There is nothing new under the sun,' is apt; there isn't any new knowledge that is unconnected to what came before. If we take into consideration the instantly backward-moving *now*, that makes the world/nature different for each of us individually, none of us is ever in a position to discover 'what the world is really like.'

It would be a mistake, though, to say that because all scientific theories fall short of absolute truth, they are all equally untrue. Because we cannot discover The Answer doesn't mean that we cannot find any answers. We do find answers, which supply what we are searching for at the time. Answers are a path toward finding other (hopefully better) answers.

The notion that there is no such thing as an objective fact that cannot be disputed, might seem quite counterintuitive at first. Surely, we can claim, there are true facts, for example, about distances and physical entities: knowledge that does not change over time. Actually, no. There are disappointments here, as well.

An example is the height of Everest. That changes over time, as colliding continental plates push up and erosion wears the mountain down. It doesn't even stay in one place, moving laterally at a rate of about 6 centimeters a year. Was our previous knowledge wrong? No, it wasn't. What we have is an example of the fluidity of knowledge, forever changing, as does everything else around us and within us.

In the previous chapter, we mentioned that 15,200 km separates the USA and Australia, as it did a hundred years ago. Shamefully, we have to admit that this information was 'true' only approximately. Not only is it a very rough estimate, but even if we were to measure from specific points on both sides, the land touching the oceans has eroded and changed tremendously over the past hundred years, and is doing so even as you read this.

Then there is knowledge consisting of facts that we know have changed and will change in the future. Samuel Arbesman, in a book whose title sums it all up nicely—*The Half-Life of Facts: Why Everything We Know Has an Expiration Date*—calls the facts that make up that knowledge 'mesofacts': facts that change slowly and continuously. So, for example, we know what the population of the Earth is; yet, we are very aware that it is different to what we knew a few years ago and that it will be different in the future. These are facts that change before our very eyes.

Let's take another breather and make some order within what we call knowledge. For a start, we can see that there are different types of knowledge that consist of:

- facts that are accepted as true (the iron content of spinach, or that the Earth was known to be flat) at the time
- facts—mesofacts—that we know are temporary (the population of the Earth, the price of oil). Here we can add facts that we know to be true, but cannot absolutely rule out changes; for example, the capitals of counties (Brasilia is the capital of Brazil, but before 1960, the capital was Rio de Janeiro)
- facts that are approximate, but are close enough to be considered true (the height of Everest, the distance from the USA to Australia)

As seemingly different as these types of knowledge are, they have a common denominator: memory. For whatever is known is brought up from memory.

Knowledge, then, of whatever type, is memory. But we can see that it is more than that. It is also a projection of what knowledge will be. Knowledge would be useless if we didn't think it would be true tomorrow. And that, of course, puts us in a bind when we accept that past knowledge was different and changed later.

With memory as slippery and unstable as it is, it's no wonder that knowledge changes over time. We can assume only that knowledge is approximate. How could it be otherwise? A continuously moving-to-memory *now* within time precludes rock-steady, unassailable facts 'out there,' covered by objective knowledge about them.

Why is this view of knowledge important? It isn't just a picante fact that has nothing to do with what goes on in our day-to-day experiences. The notion that there can be objective knowledge of the Truth is a dangerous mindset, the refuge and justification of dictatorships everywhere. As Jacob Bronowski put it when he stood in the grounds of Auschwitz: "When people believe that they have absolute knowledge, with no test in reality, this is how they behave. This is what men do when they aspire to the knowledge of gods."[70] The belief that knowledge can be objective and that it is simply a matter of finding it leads to absolute power for those who believe they have found it. To err is to be human and knowledge is a human attribute. To believe that one has found the Truth means that one is immune from error and so is above human laws.

History and memory

The granddaughter of a friend of mine had a project to do for school, which was supposed to include a family story that had been handed down over the years. My friend wracked his brain and was unable to come up with anything. So he invented one, which told of his grandfather migrating to England instead of to the USA, which was originally his intended destination. The reason he landed up in England was because he had only $25 and not $50, which he would have needed to get him to the USA. This story, completely untrue, has become part of family history. A historical fact to be handed down over the generations.[71]

History takes memory and knowledge on a long and wide journey. History is, after all, collective memory—the sum of individual memories that relate a story of what is believed to have happened. Constantly growing as it is passed from person to person, from memory to memory, history is far from a collection of static tales: perhaps likened to Chinese whispers, in so many ways. So if we were in a rough ride with memory, we are in for an even bumpier time with history.

Earlier, we metaphorically described memory as a tapestry, made up of a collection of colored threads. How apt that is when applied to history, yet with multiple more threads that change color as history is constantly written and rewritten. History's tapestry is woven together by sources written later, artistic works, archaeological artefacts, scientific technology, geography, climate and myriad threads, including later retrospective influences, that are deemed to give a whole story.

Memory is by its very nature subjective and selective. Historian Tony Judt pointed out that for this reason, memory cannot substitute for history. He was wrong. History is dependent upon the person who is telling the story. As Edward Gibbon, the famous 18th century historian, wrote in an early unpublished work, "Every man of genius who writes history infuses into it, perhaps unconsciously, the character of his own spirit."

Even more damning, history is in many ways an imposed memory, an appeal to authority, with a shelf life that lasts as long as it is believed and accepted as feasible and probable. It changes over time as new information is added. How could it not change, when what we believe to be true today will change because of what we discover tomorrow? We are in a double whammy: the changing story itself, and our feelings and personality at the time we are considering the story that is being told. With *now* existing only as a constantly backward-moving stream of events and thoughts, it could never have been a base upon which a stable story is told. And, of course, history is determined by when you start the clock: When was the start of WWII? Was it September 1, 1939, when Germany invaded Poland? Or was the beginning actually a result of the Versailles Treaty of 1919, or of the First World War in 1914, or perhaps Japan's conquest of China in 1937? If every effect has a cause, how far back does one go in order to start 'a particular piece' of history? Is there such a starting point? And could history consist of pieces, unconnected to what came before and what came afterward?

History is as much a history of memory as it is of what supposedly happened in the past, and as such we can only grasp it when it has passed.

Those thoughts put a perspective onto the dictionary definition of 'history':

> *History* (from Greek *ἱστορία, historia,* meaning 'inquiry, knowledge acquired by investigation') is the study of the past, particularly how it relates to humans.[72]
> A chronological record of events, as of the life or development of a people or institution, often including an explanation of or commentary on those events. The branch of knowledge that records and analyzes past events.[73]

In the light of what we have seen about what the past could have been, these definitions open up a Pandora's box of questions and problems. What can we know of past events that could be objectively true? Can we hope for at least some idea of what happened?

As with much of life, we are interested in getting to the truth, or as close to it as we can. In this respect, the same problems we have with our own personal memories we have with history: perspective. Therein lies one unavoidable paradox: we are too close to the present day to have an objective view, but we are too far from the past to know what happened. Added to that is the point that the more often memory is examined and churned over, the more it is distorted. What we are often left with is creative speculation.

Being collective memory, the stories that make up history are in many aspects acts of imagination, not only for the historians who are telling them, but also for us who are hearing them. And considering that much of history is based on personal stories, we need to take into account that, as Steven Novella, a neurology professor at Yale University, says, "People are horrible historians. Human memory is a malleable subjective story that we tell ourselves." This is not helped by the erroneous estimation by people of time between events and that they tend to link time that has passed with their perceptions. For example, people shown causally related pairs of historical events—such as the USSR putting Sputnik in orbit and America's landing on the moon—underestimated the time between the events by a third. Meanwhile, they overestimated the time between unrelated events, like Sputnik and Woodstock, by half.[74]

As with personal memories, history can be manipulated, consciously or unconsciously. Which events are included and how they are interpreted, which are excluded and how *that* is interpreted, what is emphasized and how they are nuanced, all these and more is what makes any story interesting along, and crisscrossing, the boundaries of truth and facts.

It isn't necessary to wait for centuries, years or days for history to change. The instant movement of *now* into memory makes it happen in a flash. "The life of a fact is short, very short indeed: a fact gets processed within seconds of its happening and then enters the machine of our mind. In no time it becomes something for me and something else for my neighbour, especially if you and your neighbour are opponents in a dispute."[75]

Sometimes we are faced with past events that did not occur but that give us useful subsequential facts. So it is with the apocryphal story of Newton and the apple. Newton's definitive biographer, Richard Westfall, does not like that story, and warns us how these populist fables can end up giving a distorted view of science. But did it give us a distorted view of science? Surely the image of Newton being hit on the head illustrates nicely the ubiquitous nature of gravity. Or take the archetypal experiment of Galileo's legendary (and probably apocryphal) dropping of weights from the Leaning Tower of Pisa. Whether or not it happened, we have, like Newton and the apple, obviously made-up events that lead us to practical knowledge.

History is generally assumed to be the past informing us in the present, and that the present has evolved from the past. As we have seen, though, time and memory play strange tricks on the past, so that history is equally as much a matter of the past evolving from events that came later.

The notion that history changes through what we are today is not as audacious as it sounds; it makes eminent sense within an understanding of *the present* as a back-to-memory moving continuum in time, with recent memories changing earlier ones.

Indeed, the notion that there exists a 'now' moment in time would make history—an account of what happened in the past—problematic. If reality is considered by *nowists* as consisting only of everything that is happening right *now*, and that *now* is all there is, by the same token the past would consist of *then* moments: freeze-frame images of only what there was, with no relevance to what happened before or after. And since each person's *now* is subjective and different from everybody else's, everyone's *then* would have been different as well, resulting in an impossibly massive off-beam record of what supposedly happened.

And why would one trust history if it is *the present* that is exclusively all-important? Presumably, *nowists* might not be pleased to know that their views are not too dissimilar to Henry Ford's: "History is more or less bunk. It's tradition. We don't want tradition. We want to live in the present and the only history that is worth a tinker's dam is the history we made today."[76]

We need to be careful, though. It's not a matter of history being wrong or false, but whether or not it is at all possible for it to relate to what supposedly actually happened. It is, after all, an account by historians of what is remembered by others. As such, it is as 'true' as is a memory. What we know about prehistory, the period before recorded events, is mainly through archaeological discoveries. Back in those days of oral tradition, a story was only what you could remember of it. And it is the memories of people thousands of years ago that historians write down, setting it all out as a record of what is surmised as actually having happened.

And how could we not mention the first orderly narratives set out for posterity in a systematic fashion? The 5th-century Greek writer Herodotus, considered the world's first historian, recorded centuries of wars between the Greeks and the Persian Empire in the early 5th century BC, relying on oral accounts that he collected. Although they included gossip, myths and rumors, sometimes providing conflicting accounts of events and inviting the reader to decide which to believe, the significance of officially recording events was enormous. Then there was the Athenian political philosopher Thucydides, known as the first scientific historian, whose history of the Peloponnesian War in the late 5th century BC was based on what was considered objective criteria, such as interviewing people who had actually fought in the war. This did not stop him from imbuing his accounts with his own political view or using his creative imagination to report lengthy formal speeches that, as he admitted, contained passages that he thought should have been said, and sometimes even inventing the speeches themselves. To complicate things further, much of what is known about Thucydides is from later sources, who had their own ax to grind.

History changes constantly, often through new findings, a long-buried artifact or a long-lost document, thereby altering the way we look at a certain civilization and its history. It happens all the time. It's enough to look at dinosaurs and how they are becoming so very different from the way we always imagined them to be; with fossils being reinterpreted through new techniques, many of us have trouble recognizing even our childhood favorites. Instead of the cold-blooded, slow-witted lizards we remember being told about, scientists think that many dinosaurs were warm-blooded and fast-moving creatures, some even with feathers, and colored patterns that included stripes and orange tufts. But don't hold onto that too tightly, since the chances are the dinosaurs will change yet again. And again.

From what we have looked at so far, it should not be too controversial to say that there is no past until it comes to us as a memory. And when it does, that memory may be of an uncertain past, or even a false one. We don't know; we can't know. Often, our memories are based on what we come across that we take as evidence of what had been, on which we build a story of the past. And let's face it: the fact that our view of the past changes constantly when new evidence comes to light, indicates how tenuously theoretical the past is and the possibility of an objective 'real' past.

Building up a past is problematic, not only because it may be partly (or wholly) fictitious, but also because it is based on memories that are theoretical, especially when they are of events that occurred when there was no one around at the time. At the period before prehistory, then, there is a problem. There were no humans in existence at the time of the dinosaurs, yet we have built up a memory of what they looked like, what they ate, their habits and so on. Whenever a dinosaur bone is unearthed, we add to our memory of the times—but to no one's memory of the time: in fact, a false memory. It cannot be anything other than a false memory, since it is built on conjecture based on evidence that experts have put together later.

It should be clear that the issue of memory in this context is not meant as disparagement of paleontology as a science and the work that is done on forming a possible picture of what was in existence before the advent of humans. Some of the findings—for example, that evidence of webbed feet indicate an aquatic life—add a reasonable addition to our memory of the past. Our discussion is about the nature of knowledge and the vicissitudes of memory on which it is based.

It's the question of knowledge again: with history considered to have been correct before later discoveries showed it to have been different, it should make us wonder about what we consider to be correct today, when it is likely that new finds will be made in the future.

It's all rather worrying. The more we look back into history, the more we must wonder how much, if anything, was factually recorded. Most often, history is an attempt to attest to how things might have been, since we simply don't know what actually was.

In Selinunte, the 7th-century BC Greek city in Sicily, slightly away from the main Greek temples is a structure consisting of four columns topped by a massive construction. When visiting, I was amazed to hear that it was constructed from columns and parts of columns, to which were added parts of the top, all of which were found strewn around nearby. Archeologists had constructed it, much like a Lego set, in the way that they, as experts, thought it had once been. My amazement was due to my naiveté. It might originally not have looked like that at all; the columns may have been from different buildings. It was a creative justified guess, but not necessarily what had actually been. Science archeologist Walter Alvarez had a point when he said that "Understanding how we decipher a great historical event written in the book of rocks may be as interesting as the event itself." It would be interesting to know what archeologists 2500 years from now would make of the ruins of some of our well-known buildings, say of Trump Tower in Manhattan, New York, with its gold furnishings, or the pyramidal Shard of Glass in London, UK, or the Eiffel Tower in Paris, France. Would they be considered as temples, thrusting upward in homage to gods? What would they look like after reconstruction to show their original shape and function?

Once we begin to look at some of the most well-known archeological sites, we are faced with constructions that are at best faithful reconstructions of what fell to ruin, or at worst like those in Selinunte—constructions of what historians believe may have been there previously. One of the most audacious reconstructions is at Knossos, the grand and flamboyant Minoan palace in Crete, that can charitably be said to be as much the creative imagination of the British archeologist Arthur Evans, who excavated the site from 1900, than to what might actually have been there during the Minoan period of 1900 BC. Who knows what The Grand Stairway looked like at the time, after Evans reconstructed it along the lines of an Italian renaissance stairway 3500 years later!

With all the difficulties there are in reconstructing what was by restoring a fresco or a Roman wall, it is even worse when it comes to an ephemeral profession, such as music, where there is an ongoing discussion as to the merits of rendering works in an accurately historical form or bringing them more up to date. But replicating early music compositions is not feasible, since there is nothing to build upon other than the musical notes. Performance of classical music cannot be considered as an archeological way to perform music; it is, in fact, just an artistic hypothesis. Music doesn't exist any longer once it is played: it is just air that vibrates. We don't really have the music of the past. We just have indications from historical sources—sometimes with contradictions between them—and we have scores. But scores themselves don't make any sound. Both music and sources need the mind and inspiration of the performer to be converted into sounds. So what are now set pieces, even with the leeway of interpretation, are reconstructions, in some way not that different from accepted archeological ones.

One of the world's most iconic structures is the Parthenon in Athens, Greece. Built as a Temple to Athena in the 5th century BC, it was successively a church and a mosque (with added minaret) before being blown apart by an exploding shell in 1687. The structure we see today has been reconstructed several times, and although not the original stone, it is believed to be a fair copy of what had been originally. What is not generally known (possibly owing to Greek pride in the icon that reflects the country's ancient past) is that the periods when the structure was a church and a mosque were very much longer than its use as a Greek temple. In fact, from the time that Greece won its independence from Turkey in 1832, the aim has been to strip away all evidence of its non-Greek past and to restore it to its Golden Age, resulting in a restoration in which most of the structure has been taken apart and put together again.

This brings us to an intriguing question, one that has been struggled over for quite a while. Known as the Theseus paradox, it considers whether an object—the original example being the mythical Ancient Greek ship *Theseus*—remains the same object if its parts are replaced. Was it still the ship *Theseus*, when all its parts were gradually eroded and changed? Coming to a more contemporary, but similar, example, is HMS *Victory*, Lord Nelson's flagship at the Battle of Trafalgar in 1805, of which only 17 percent of the original timbers are still in place. Similarly, the *Cutty Sark*, the last British tea clipper to be built and one of the fastest, was badly damaged by fire in 2007 and is now almost a complete replica. In the USA, USS *Constitution* (1795) has had most of the ship's timber replaced over time, with only 10–15 percent of the original remaining. What is it that gives those objects their identity? Any object, in fact. Is it its physical form, its name, its spirit, its character—all of those or a combination of them?

So far we have been considering the notion of identity *despite* change. But we can see the opposite in Japanese culture, where stability is brought about through *deliberate* change, where it is common to destroy shrines and rebuild them. Every 20 years, the most sacred Shinto site in Japan—the Grand Shrine at Ise—is completely torn down and replaced with a replica, constructed to look as weathered and authentic as the original structure built by an emperor in the 7th century. For the Japanese, then, identity is the spirit, not the physical. This is applied not only to shrines, but also to the destruction of some beautiful buildings that would, in the West, be considered as iconic symbols. The Hotel Okura, an icon of Japanese modernism built in 1962 to commemorate the country's hosting of the 1964 Olympics, and cherished for its unique and atmospheric lobby, has been demolished in favor of two glass towers, meant as symbols of modernity to coincide with Japan's hosting of the 2020 Olympics. Where other countries keep structures built for their Olympic Games, Japan destroys hers in order to build new ones, so that the once state-of-the-art National Olympic Stadium, designed by Mitsuo Katayama for the 1964 event, is being replaced by a new structure to celebrate tomorrow's idea of futurism.

Our original reference was to the Parthenon in Athens and the history of that edifice. But we can apply the conundrum of *Theseus'* ship to memories and who we are—our personal identity. If we are the sum of our memories and experiences, what happens to our identity if our memories are eroded, lost, or are not real memories at all? After all, we have our own idea of who we are, as do other people. How do we keep on recognizing ourselves if our memories change over time? And what does it say about our identity all together if we realize that our memories might not reflect what actually happened? At the most extreme is amnesia: In which way is a person with no memory the same person as they were? When we say 'John is not himself,' we mean that he is not like he was previously, not that he is not John.

Identity is an essential part of the collective memory of history, so much so that the very search for the past can turn out to be as destructive as it is obsessive. A classic example is that of

Nazi Germany, which made the superiority of a Germanic race a basis for the legitimacy of its cause. In 1935, Heinrich Himmler founded an organization called the Ahnenerbe ('Ancestral Heritage'), whose claimed goal was to discover "new evidence of the accomplishments and deeds of Germanic ancestors using exact scientific methods."[79] These included projects to find the lost city of Atlantis, the Holy Grail and an expedition to find the origins of a supposed Aryan ('Indo-Germanic') master race in Tibet. As crazy as that sounds today, it was taken seriously as an attempt to write into memory an ancient civilization from which the Nazis believed the Germans had descended.

History and human memory are, as we can see, intimately related. Attempting to look at the past is an attempt at understanding humankind. History, after all, is a study of the past, as it relates to humans and how humans relate to history.

It's all a matter of time, of course. It's not surprising that history as 'facts' is so slippery. For how can it not be, when time itself is change that has already occurred? Since whatever we examine is already in the past, we are witness only to what was. This being the case, *Theseus'* ship is not a paradox. It is, rather, a reflection of the inevitable backward movement of the so-called present. In that sense, nobody ever saw *Theseus'* ship in the present, even when it had just docked. In the same way that a person's cells are constantly being replaced by new ones, yet keeping his identity—his Self—so it was with *Theseus'* ship, which was constantly being changed yet kept its identity.

History consists of stories told by humans for humans, based on memories of other humans, that consciously or subconsciously reflect personal preferences and interpretations of events. So as far as *Theseus'* ship is concerned, it stayed as such even when there was nothing left of the original, because people were interested in perpetuating the story. And that is basically how history works: stories remain real because there is an interest in keeping them so, perhaps political or national. Not only that, but they cease to be history when there is no reason to perpetuate the story.

History, like memory, does not necessarily differentiate between the important and the trivial, which are so in the eyes of the beholder at the time. It has no way of doing so, largely because past events are not only relative, but also completely subjective, and depend on context at the time and when related. Where will future historians place a football rulebook, *The 1863 Football Association Minute Book*—which, in 2015, was considered so important as to be placed alongside the *Magna Carta* in the British Library as part of its 150 anniversary celebrations? If, as announced, the book "changed the world," has it done so in the magnitude of a document from 1215,[78] which established for the first time the principle that everybody, including the king, was subject to the law and hence was the first spark toward eventual democracy several hundred years later?

Important or not, as we have seen, the further back the memory, the more it is chiseled into reality. An examination of a few of those memories make us wonder. Let us look at the kilt, an article of apparel as Scottish as can be imagined. But is it? According to research done by the late historian Hugh Trevor-Roper,[79] the kilt did not exist before the 18th century; nor did the clans distinguish themselves by the pattern of their tartans until they were taught to do so by an English clothing manufacturer. Even more shocking is Trevor-Roper's contention that Scottish history altogether is an invention, very largely by Hector Boece, a 15th-century scholar and humanist from Dundee, who wrote a hugely influential *History of the Scots*.

It's worthwhile lingering some more on this particular point, to demonstrate how memory can possibly be implanted and made to stick by self-perpetuating repetition that deepens belief in its reality.

Boece, ignoring the origins of the Scots as an Irish people who migrated to western Scotland in the 5th century CE (*'Scotus,'* in

Latin, originally meant 'Irish'), elaborated a myth, according to which the Scots were descended from an Ancient Greek hero, Gaythelos, and his wife, Scota, an Egyptian princess who was the daughter of the biblical pharaoh. This was to rival the antiquity of the English, who traced their descent to Aeneas and the Trojan Wars. The problem Boece now had was a 900-year gap in the historical record. What did he do? According to Trevor-Roper, Boece simply invented 40 kings, who he not only named, but also supplied with detailed biographies, so creating a national identity.

The circularity of time within history is complete when the recounting of stories that influence our perception of the past eventually affect later events. The mythical history of the 40 kings was used to legitimize the deposing of Queen Mary in 1567 and convincing Queen Elizabeth of England that the Scottish constitution had always allowed for the removal of kings. According to Trevor-Roper, this worked, despite the fact that the entire period of the 40 kings had been invented.

It is not surprising that criticism of Trever-Roper's views came fast and furious, especially in the mounting resurgence of Scottish pride and separateness. But the point, and the reason that it has been included here, is not about Scotland and whether its history is true or not, but about the shifting tides of stories and memories of them that are implanted into consciousness. What, after all, are national flags, national anthems and state symbols if not planned, implanted memories? And within plain sight are the solid manifestations, the statues of national personalities, that are particularly vulnerable as to their place within historical memory. So it was with the statue of Cecil Rhodes, the British mining magnate and African colonizer, pulled down from its plinth at the University of Cape Town in 2015, with a later demand by students at Oxford University for Rhodes' statue to be removed also from there. Tampering with history is tampering with memory, a sensitive and explosive issue as illustrated by the statues of Confederate heroes that are the focus of periodic convulsive demonstrations in the USA South. The violence and loss of life in Charlottesville, Virginia, in 2017 over the statue of the Confederate general Robert E. Lee was emblematic of a struggle to perpetuate a historical memory versus the urge to lose that memory. Heroes, it seems, are in the eyes of the beholder's memory.

Nothing is sacred, it seems. Even an immutable concept such as God has changed over time, not only from religion to religion, in the monotheistic religions from a common source in Judaism through Christianity and Islam, but also within each religion itself. After all, the story of God is part of the story of humans, in which he has been adapted to varying cultural circumstances (even becoming 'she' for followers of feminism and fans of political correctness). Hardly recognizable is the spiritual God of the desert to the father-like God of today, who nevertheless assumes many different forms, even among believers.[80]

We don't need to go that far back in order to see how history can be changed, for we can see it happening before our very eyes. In June 2004, Germany was invited for the first time to the D-Day commemoration at the invasion beaches in France. German chancellor Gerhard Schröder used the opportunity to align Germany with the Allies, telling his audience that D-Day was not a "victory over Germany, but a victory for Germany that led to its liberation from Nazi rule." This touchy-feely version of revisionist history implants a false memory. Here was a deft change of national memory, rewriting history to give victory for both sides. In one or two generations, it will no doubt be fact that Germany won WWII together with the Allies.

Myth and fact become inexorably intermeshed, since they are often part of what shapes nations' characters and self-perception. The Great War of 1914–1918 with its insane slaughter brought out myths galore, based on genuine heroism. For Australians and New Zealanders, there is enormous symbolic importance to the performance of the two Anzac formations, that have become a foundation stone of perceived nationhood and importance of the two countries as independent entities. Almost universally accepted is the blame placed on the British, who sent the Anzacs on a misconceived invasion that turned into months of slaughter, including the iconic butchering by the Turks at Gallipoli. However, nearly twice as many British soldiers perished in the doomed campaign as did Anzacs: 21,000 against 11,430. As folk memory of Gallipoli, those figures aren't part of public knowledge.

It has been said that history is written by the victors. That, of course, is only partly true. It would be more correct to say it is told in a way that the historian wants it to be known. A classic example is the history of the American Civil War, where the official version, written by victors (Federal North), is juxtaposed just as prominently by a version written by the losers, the Confederate South. Not just the 1861–1865 war itself, but also the events leading up to it and the reasons for it. It is not surprising that 'truth' and 'facts' lose any meaning within a wish to perpetuate a convenient history. So it is with the death of General Albert Sidney Johnston of the Confederate army (and the highest-ranking soldier killed in that war) in the terrible bloody Battle of Shiloh in 1862. The point here is not how he was killed (although even that is not as clear as set out in the perpetuated version), but the symbolic tree stump that marks the exact spot where he was supposedly killed. As a place of pilgrimage, it has taken on proportions of a hallowed icon. The fly in the ointment is that tests have shown that that stump was not of a tree that would have been in existence at the time. But history is history, even when it isn't. Protests stopped the intended uprooting of the stump, which remains to this day as a relic of false history, to be worshipped on the altar of a misremembered past.[81]

Like memory, history is selective, sometimes subconsciously, but possibly just as often, consciously. The killing of unarmed prisoners is always done by the other side, with evidence as to atrocities happening on 'our side' generally not available to historical memory.[82]

When a historian writes: "Social memory is essentially dynamic. It grows and changes, constantly recovering or burying facts and meanings. Historians are part of this process: our work provides elements to allow individuals to feel part of a larger, collective whole,"[83] this is a hugely problematic statement. It says less about 'historical truth' than it does about the elasticity of history and the possible manipulation by historians. If history is merely a common denominator of changing facts, done so that individuals can feel part of a collective whole, we will have moved away from what we expect history to be. Perhaps, though, that is what history is after all, reflecting the dynamic constantly moving '*now*.'

Collective memory, like individual memory, can disappear, as happened when the Alexandria Library was burnt on a number of occasions in the ancient world. The repository of thousands upon thousands of scrolls were burnt and disappeared, just as previously the tablets of the Assyrians had disappeared beneath the hills that overlook the Tigris and Euphrates. The Romans would follow their practice of library building. But writing, for all its powers of memory, crumbles again and again at the implacable forces of time, war and change. Gone, too, are the ideas set out in countless documents that we don't even know were ever around, since we cannot know what we have forgotten. It is only through the quotations and references made by other ancient authors, and by their summaries of his ideas, that we know about the lost books of Democritus, who, in the 5th century BC, wrote about physics, philosophy, ethics, politics and cosmology, the nature of language, religion, the origins of human societies and on much more.

In many ways, history is as much about what has been forgotten as has been remembered—as much as what did not happen as did happen. Rather than being unchartered territory, it is possibly uncharterable territory.

Making history

Perhaps not surprisingly, history is to a large extent a clash between nostalgia and an extremely murky past. Nostalgia is so fuzzily wonderful because what has been remembered is enhanced by what has been forgotten. History can be good at leaving out uncomfortable details, the problem being that we cannot know what they might have been.

The notion that what happened was better than what is happening is an extension of the all-so-common memory of summers of our youth having been longer, hotter, more carefree and generally more fun. As John O'Connell notes in his book, *For the Love of Letters*, the past is generally perceived to be better than the present, whatever the present holds. This has a long history: Socrates is quoted as stating: "The children now love luxury; they have bad manners, contempt for authority; they show disrespect for elders and love chatter in place of exercise."[84] In the Woody Allen's 2011 movie *Midnight in Paris*, the character Gil Pender is transported every evening to the 1920s, where he meets Adriana, and both are eventually transported back to the 1890s Belle Époque, an era Adriana considers Paris's Golden Age. It is when Gil hears Henri de Toulouse-Lautrec, Paul Gauguin and Edgar Degas state that the best period was the Renaissance, that he comes face to face with the realization that people are never satisfied with the present and always consider the past to have been better.

Accessing old memories and reassessing them seems to be a very human thing. What we now love about Paris, Baudelaire in 1855 decried as over-urbanization and senseless modernization. 'What is the world coming to?' has been a plaintive cry throughout history. Down to earth was Tony Hancock's "I don't know whether it's possible to be nostalgic for a time that didn't exist, but I think I am."[85]

Every generation has gone on about the good old days. The Ancient Greeks and Romans fretted about the young and their callous disregard for the hard-won wisdom of their elders. The 17th and 18th centuries had their poets, writers and philosophers who famously idealized the Noble Savage, a being who lived in harmony with nature and did not destroy his surroundings. Nowadays, we worry about our kids as 'digital slaves,' who grow up surrounded by electronics, and can't put down their smartphones and iPads. No doubt our Paleolithic ancestors felt something similar when they painted images on caves to show what was outside, while yearning for the good old days of plentiful food.

In that light, prehistory was best of all, of course! A world in which everyone is imagined to have lived in unspoiled nature, where people were content with their lot in social units all living in harmony, and natural medicines were bountiful and effective. The corollary is that our modern lives are out of touch with the way human beings evolved and that we need to redress the imbalance, as advocated in newspaper articles, on morning TV, in dozens of books, and by self-help campaigners who promote slow-food or no-cook diets, barefoot running, sleeping with our infants and other measures to get us to live more like our ancestors.

Vacations that take us to constructed primitive jungle villages (with all the mod cons) are a way of constructing a history, a memory, of what may not have existed, reinforcing what wasn't.

Wishful thinking about a perfect past (and no doubt an idealistic utopia just over the horizon) is one thing; digging up what might have happened brings us down to earth. Prehistoric man may not have been as gentle and loving as we like to imagine, as recent findings have shown. Researchers at the University of Mainz, Germany, studying the remains of 26 people dumped in a pit dating around 4000 BC, have shown evidence of needless violence, of torture and mutilation both before and after death. The remains, taken from a war grave in Darmstadt, Germany, dated at around 5000 BC, demonstrate a pattern of large-scale violence. And it seems to have been common long before that, when humans were still foragers and hunters, as was unearthed at Nataruk near Lake Turkana in Kenya, where the fossilized bones of a group of 27 hunter-gatherers, who were murdered 10,000 years ago, was discovered. Four victims, including one heavily pregnant woman, were bound by the hands and feet before they were slaughtered. The others showed signs of extreme violence, and some had blades and arrows still buried in their skulls.[86]

Not unlike more recent history, massacres in prehistory were a strategy, as were acts of mass violence and systematic torture, which were repeatedly carried out in organized warfare.

The fuzzy edge of nostalgia is everywhere. We can see it alive and well in the books of Jane Austen with the idea of a gentile England, where even the poor people didn't have it so bad, really. Reality was very different, though. In the world outside, children worked the mines, people dug up freshly buried corpses to supply body parts for research, while still in 1816, a year before she died, human excrement floated down the streets of London. When Austen was fourteen, in 1789, the last woman was burnt at the stake in England.[87] While Austen was writing, the Napoleonic Wars were raging. She had brothers in the navy, and saw ample evidence of economic and emotional disaster. Most pithily apt perhaps would be the observation of conservative cultural critic Ernest van den Haag in 1956, that "There have been periods happier and others more desperate than ours, but we don't know which."

Nostalgia plays havoc with even recent memory. East Germans wax nostalgic about the good old days of the Democratic Republic of Germany (1949–1990), buying things they hated in the past, with specialty shops and some websites offering East German board games, 15-packs of the infamously rough Cabinet cigarettes, Be Ready condoms and even cans of Trabi Duft—fumes from the iconic Trabant car. Many Russians now think back with longing to the days of the Soviet Union, as do some now-independent republics that had once been part of the Soviet Union. Similarly, the days of Franco's Spain are to some extent looked back on with nostalgia, where memory is not of repression and lack of freedom, but of order and security that is seen as lacking in the anything-goes world of unemployment, open pornography, tawdriness and trash. And in Italy under Mussolini, 'the trains ran on time'!

We can't get enough of nostalgia, with TV sending us back in time with its portrayal of aristocracy and the gentry, that has been called the 'warm bath' feeling of the reassurance of an ordered world, where everything is as it should be. This was exemplified in the series *Downton Abbey*, which was such a success on both sides of the Atlantic—in fact, Britain's most successful TV export ever.

Nostalgia makes history, and history feeds the nostalgia. Atlanta, Georgia, is home not only to the real thing, according to a slogan of Coca Cola, but also to a nonexistent reality, that tourists take seriously when they look for traces of the lives of Scarlett O'Hara and Rhett Butler within their story of the Civil War that was made so real in the movie *Gone with the Wind*. As Tony Horwitz[88] tells us, the movie had done more to keep the Civil War alive, and mold its memory, than any history book or monument. The fact is, the movie wasn't even filmed in Georgia but in Hollywood; although Atlanta makes up for it by having a museum of movie memorabilia to keep tourists satisfied and, incidentally, to reinforce that particular myth of historical reality.

Nostalgia was always around, but today it may be text and email overload that gives the past an extra boost and longing that sends people to period drama for a simpler, slower life: a false memory, in fact, of what is imagined as a simpler past.

Nostalgia imprints false memory that softens the hard edges of a time that was anything but simple and pleasant. Millions of soldiers died in the 1914–18 war, a quarter of a million died of Spanish flu shortly afterward, not to mention infant mortality, terrible dentistry, no heating, wretched living conditions for the majority of people, and certain death from cancer and a variety of other diseases, many of them caused by London's ubiquitous fog: more than 11,000 Londoners died of bronchitis in 1886 alone. In November 1948, fog resulted in around 800 deaths, while four years later, a 5-day fog was directly responsible for about 4000 deaths.[89] So much for 'the good old days.'

Perhaps it is not surprising that recording collective memory is prone to schizophrenic waves of guilt. With collective memory being changed to accommodate that which seems appropriate for *the present*, the other side of nostalgia is ultimately an urge for what we now call political correctness.

Nowhere is this more revealing than in recent history books on the founding of the USA, which seem have swapped an attempt at telling what happened with some sort of moral hand-wringing. The issues of white privilege in American society, and the inequalities of race and gender, have seriously affected the writing of American history, and have become an agenda to denounce the values of the past in order to assuage a collective conscience. Using history to deal with current problems is a suspect proposition. Herbert Butterfield puts it well: "The dispensing of moral judgments upon people or upon actions in retrospect, is the most useless and unproductive of all forms of reflection,"[90] while Simon Sebag Montefiore calls it "the temptation to see the past through the obsessions of the present";[91] although it supports those apt Talmudic words: "We do not see things as they are, we see them as we are."[92]

There is a thin, fuzzy line between learning from the past and rewriting it. By remaking history according to, and influenced by, our present perspective, we actually risk sabotaging our aims. We have created a double-edged sword, in fact, since not only do we get a slanted view of what happened, but our later views are built upon, and compared with, a romanticized past. Robert Kennedy put it like this: "There are people in every time and every land who want to stop history in its tracks. They fear the future, mistrust the present and invoke the security of a comfortable past, which in fact never existed."

History is, in fact, often a series of implanted memories. On a daily basis newspaper editors decide which news and photos are published out of the countless ones that are snapped. The power of the editor, who can decide on how we will remember our past!

Even more influential, because it gets to more people, is the news that is constantly streamed onto social media from myriad uncontrolled and untrustworthy sources. Suddenly, the term 'fake news' becomes all the rage—so much so that the issue became news itself—at around the US presidential elections of 2016. Since no one really knew which news was fake, which wasn't or which was partly so (and in any case depending upon what one wished to believe), it all dripped into the collective memory of what was happening. All of it was news, in fact, with it all appearing indiscriminately on Google and search engines, where it remains. And so history is made, with the passing of time blurring the boundaries between what was believed by whom and for what reasons. All that will be left for future historians is a mass of informational memory upon which theories of what happened will abound.

Fake news in various forms, even if usually centrally controlled, has always been part of what eventually became part of history. Within this context we can understand what Berthold Brecht meant in 1931, when he wrote, "The camera is just as capable of lying as the typewriter." It is instructive to see how presidents and prime ministers of countries, whether democratic or not, are positioned in the center of a group and are photographed from angles that make them seem taller. If we consider, as well, the fact that news broadcasts of an item of news vary in different countries, subjective memory in the making is clear.

An interesting incident illustrates this occurred in 2010, when the Egyptian newspapers showed then-president Hosni Mubarak with world leaders walking together at an international conference. In that picture Mubarak was in the front, whereas the original photo, shown everywhere else, had Mubarak at the back. National pride would not allow the leader to be shown at the back and so the photo was doctored to give the desired result, inculcating historical consciousness.

Changing the past, sometimes erasing it completely, is far from being unusual. In the Soviet Union, airbrushing leaders from photos and erasing their names from earlier news reports was par for the course. This is prevalent in repressive regimes, such as North Korea, where in November 2013 Jang Song Thaek, the influential uncle of Kim Jong-un, was 'purged,' disappearing physically and historically from official memory. Matters became even worse when, reminiscent of George Orwell's *1984* Ministry of Truth, 98 percent of the country's official history was purged from its official archives.

Destroying evidence of previous civilizations has been part of warfare for thousands of years. Not that long ago, in World War II, German armies destroyed art and historical buildings in order to eradicate Slavic culture, plundering anything valuable that was considered suitably 'Germanic.' This was at the same time as the Nazi German regime was methodically not only wiping out the Jews, but also burning books written by Jews, and destroying synagogues and Jewish artefacts, as a means of wiping out traces of Jewish cultural heritage.

It's still going on. Increasingly, monuments and cultural artefacts are not simply damaged in the conduct of war, but are instead deliberately targeted for erasure as part of campaigns of ethnic cleansing or for other iconoclastic reasons. At Bamiyan, Aghanistan, three colossal statues from the 5th century—one of them a 53-meter high-standing statue of Buddha, the world's tallest, carved on the cliff face of a mountain—were destroyed by the Taliban in March 2001, on the grounds that they were an affront to Islam.

The wiping of memory is equivalent to the wiping of existence, it seems. Behind the ISIS rampage—which includes the deliberate destruction of Ancient Roman, Christian and Shia Muslim sites in Iraq, Syria and everywhere else it could get to—is the fervent belief that wiping out memory is as effective as wiping out people. There is nothing new here. Early Christian fanatics destroyed Greek and Roman art, while within itself Christianity—when split in the Reformation in the 15th century—was busy destroying art: the Protestants destroying Catholic art and statues, and the Counter-Reformation in the 16th and 17th centuries seeing Catholics attacking all things Protestant.

Wiping out a people's past in order to legitimize the genuineness of the latest present—delegitimizing the other in order to legitimize oneself—is widespread. It is a relentless and conscious changing of collective memory—an ongoing example being the denial by Palestinians of a past Jewish presence in Jerusalem and the Holy Land, while national museums are prime examples of changing memory to fit the national narrative. In the museum of Gdansk in Poland, for example, there is but a tiny hint that the city had once been Danzig and part of Germany.

Purposeful forgetfulness can be very effective, indeed. Until the recent digging up of a Viking haul of coins in 2015, the existence of Ceolwulf II of Mercia within early English history had been largely forgotten, having been airbrushed by Alfred the Great, a rival king, 1200 years ago. Only briefly described in the Anglo-Saxon Chronicle as an "unwise King's thane" (non-legitimate king), Ceolwulf II has now re-jogged the collective memory and prompted the rewriting of English history.

The liberal Tippexing of historical memory is so common that it may be the only thing we can say for certain about history! In UK theater director Emma Rice's play *The Empress* she shows how many black and Asian people lived in England in Victorian times. "Isn't it interesting," she says, that "you never see a black face in our TV adaptations of Dickens. It was the same with the Second World War—we had thousands and thousands of black servicemen, not only the American GIs but also from the colonies." Also not talked about at the time was the fact that thousands of black soldiers fought for the Union in the American Civil War and later in the First World War, while in the Second World War over 2.5 million African-American men registered for the draft, with black women volunteering in large numbers, too.

We should, of course, not be surprised by the unreliability of the stories we are told of our past. How big, though, is the step between that and 'what-if' history? Generally known as counterfactual history, it does more than falsifying what actually happened; it changes the whole basis of the events. What if Kennedy had not been assassinated? What if Germany and Japan had won WWII? While this is more fiction than history, it has the effect of querying a possible future (what if?) through shaking implications of the past.

And where on the scale of rewriting of history are 'truthers,' with their conspiracy theories that they apply for almost every event? Anything, from denying the Holocaust, to Kennedy's assassination on November 22, 1963, to the destruction of the Twin Towers on 9/11, 2001, is grist to the mill of those who turn our memories upside down by information over which we have no control. It makes very little difference what evidence and proof are presented. Although a small minority, 'truthers' take advantage of the frailties and imperfections of our memory system that are part of what makes us human.

The feeling that anything *can* happen does not mean that everything *will* happen. While it is tempting to think, as does

British writer Howard Jacobson, that "anything is possible, history has taught us that,"[93] it is a dubious notion, as is the contention that history tends to repeat itself. History has shown us what possibly *was*; we can only conjecture as to what is still possible. If history has taught us anything, it is that what will be, may be a variation of what *was*. Putting it succinctly, the future is anticipated memory.

There are no events that can be exactly the same and that also applies to past events. In the sense that it is natural to compare events, some will seem similar. Yet, no events can be the same by the very virtue of having taken place at different times. According to *The History Manifesto*,[94] the role of historians is "Renewing the connection between past and future, and using the past to think critically about what is to come." That, though, cannot be anything but limited. Not only were circumstances inevitably different, but past events have affected and so changed us as well. We don't, then, learn from our mistakes, since what we perceive as having been mistakes occurred in circumstances that cannot be repeated and, in fact, are not repeatable. What is fairly sure is that we have changed and that every generation changes; the constantly backward-moving *now* into our memory does not give us a chance to learn from past mistakes that simply do not recur as they did within their particular context at the time. Mistakes are doomed to be made again and again, each time slightly different from previous ones. And we cannot but take into account a recurring theme of memory, made so pithily by Bernard in the television series *Westworld*: "How can you learn from your mistakes if you can't remember them?"

No part of science or humanities is static, and that is not merely a reflection of human advance in knowledge, but owing to the nature of time as change within the continuum of movement-to memory. Psychology, philosophy, art, music are all part of the context of the world, all moving in relation to each other and within themselves.

Since nothing can be the same, there isn't anything that repeats itself. An orchestral score consists of notes on the page. But the annotations on the manuscript will dictate how the music should be played according to how the conductor wishes it to sound. And even then, each performance is unique, even when performed again a day later, depending on myriad nuances and feelings of the conductor, orchestra players and their reaction to the audience.

A quote [wrongly] attributed to Albert Einstein defines insanity as "doing the same thing over and over again and expecting different results." The point is, though, that since we can never do exactly the same thing again, insanity would be expecting the *same* results!

Paradoxes, riddles and the past

The path we are taking through our look at time comes up with prizes. One of them is an easier life. Certain problems that have occupied people's minds can be disassembled with ease when we take into account the reality and importance of the past. It is refreshing to know that when we accept the nonexistence of *the present* and the supremacy of the past, we can reduce mountains to molehills by showing that hitherto difficult questions aren't difficult at all.

These next two well-known problems will be interesting to tackle: *the tree in the forest* and *Schrödinger's cat*.

The problem of the tree in the forest was raised by George Berkeley, who asked: "If a tree fell in the forest and no one heard it, did the tree make a sound?" To those not into philosophy, it's an annoying question to which the answer is intuitively that of course it made a sound. But did it? How would we know? At this point, we need to ask ourselves about the tree itself. How do we know it fell? After all, the question as to whether there was a sound is contingent on a tree actually falling. In other words, if we don't know whether a tree fell, we don't have to bother about whether or not it made a sound. We can make that point clearer by posing the question differently: If a tree fell in the forest and no one knew about it, did a tree actually fall?

Putting it that way, it is even more of an annoying question. Why? Because really it shouldn't be asked at all. It's actually a trick question, since when we query if a tree fell, we are already assuming that the tree might have fallen. Rather unfairly, the question already contains the *possibility* of the tree having fallen, thereby implying that if it fell there would be someone who knew it did.

The entire question is barren, and it is somewhat surprising that such an enormous mountain was made out of the tiniest molehill. This big question is not necessarily one for philosophy,[95] but rather, and more appropriately, one that has to do with time.

It's quite easy, really. Surely we can know if a tree fell only after the event. If no one knew that a tree had fallen, there is no reason to guess that it had, or even to think about it. So why would we ask the question? To put it succinctly, you can only know about something after it has happened.

From trees to cats.The longstanding paradox of Schrödinger's cat, devised by Austrian physicist Erwin Schrödinger in 1935, is a thought experiment designed to demonstrate a supposition in quantum mechanics that a subatomic particle can exist in multiple states *until observed*. In other words, the very fact that there is an observer can change the state of a particle. Schrödinger's idea was to apply the principle not just to non-visible minute elements, but also to everyday objects—in this case, a cat—the weird notion that something can be thought of as *both alive and dead at the same time*, and that the outcome (whether it is alive or dead) depends on the observer. The idea is for a cat to be placed in a sealed box in which there was a vial of poison, a Geiger counter, radioactive material and a hammer. The minuscule amount of radioactive material had a fifty-fifty chance of being detected over the course of an hour. If the Geiger counter detected radiation, the hammer would smash the poison, killing the cat. Until the box was opened, it was impossible to predict if the cat was alive or dead, and so it would exist as dead and alive simultaneously. It should be noted that Schrödinger himself used the cat example in order to show how ridiculous the whole idea was.

Let's be clear—we are not bringing Schrödinger's cat into our discussion in order to delve into quantum physics, but to relate to a basic point that Schrödinger makes: the possibility of knowing *only when there is an observer* whether the cat had actually been alive or dead; in other words, the very fact of there being an observer allows the cat to be either alive or dead.

Again, this is easy when it's thought of as a matter of time. From what we have seen so far, there should, in fact, not be a problem with this poor cat, since the question as to whether it is alive or dead before the box is opened is irrelevant. The question, surely, is whether the cat *was* alive or dead within the box. It could have been alive and it could have been dead. Since we cannot possibly know which it was until the box is opened, the notion that the cat was both alive and dead at the same time is nonsensical.

Both of these examples—the tree in the forest and Schrödinger's cat—have in common the question: Does observation affect outcome? But that, as we have seen, is irrelevant. It is, surely, simply a matter of knowing what happened only after the event has taken place; in other words, when it is in the past. Neither the tree in the forest nor Schrödinger's cat is, then, a paradox or a riddle.

These are sticky points of principle, though, that have a direct bearing upon our discussion about memory and the past. Both questions are attempts at getting to what 'really' happens. But this supposes that there is an actual real thing that 'is' somewhere even if it can never be observed. Is there 'really' a tree that fell if nobody knows about it? Was the cat 'really' alive or was it 'really' dead before the box was opened? Or, as we asked about memory, were there objectively real events in the past, even if we can never get to them? The prophet Isaiah didn't seem to think that this was a question worth thinking about when he said that our Earth would one day vanish, and that it would "not be remembered, nor come to mind" (Isaiah 66:15). Well, isn't that obvious? Who would be around to remember and whose mind would it not come to? If there is no way to test any statement of 'fact,' it is worthless.

And so back to paradoxes. The classic ones, and possibly the oldest, best known, are Zeno's paradoxes. Zeno, born around 490 BC, wrote a series of paradoxes to show that, despite one's observation, change was impossible and that therefore movement was an illusion. Among the paradoxes was that of the arrow, which briefly goes like this: if you examine an arrow flying through the air, at any point in time it will be still. So, says Zeno, at *any* instant, the arrow will be still. The logic is that if an arrow's flight consists of still moments, how does it move? Is motion, then, an illusion? he asks. Obviously not. But it would be, and so would time itself be an illusion, if it consisted of a sequence of 'nows.' The point, says Zeno, is that the arrow is never still, since it cannot be still and also move.

The paradox can be, and has been, solved mathematically; but that is not relevant to our discussion. Our interest is the paradox and its connection to time. But there is, in fact, no paradox here, since time is not split into 'nows,' and nor is it composed of a series of events that could be stopped and examined. As already mentioned, time is not like a movie, which can be stopped in order to look at a still. The movie will have stopped, but time will continue no matter what.

Time, unlike a movie, is a continuum and doesn't stop, however much we would like it to, sometimes. As with all events, what we see as the movement of the arrow through the air is fractionally already in the past. We can see change only after it has happened. Time, the result of continuous events within a continuum, all of which have already passed, is not open to examination. There is, in fact, no paradox, since the movement of the arrow has already happened when we think we see it happening in *the present*.

So while we can say that time is change, that is not the full story: because there is no *now*, time should be seen as the result of change.

That brings us neatly to another seeming paradox: we can't see time passing, even when we know it is.

If you were to stare at a plant, you wouldn't see the plant changing, no matter how long you stared at it. Has time stood still? Of course not. Time continues on its merry way, our watches showing it and the second hand moving onward. The plant is obviously growing or changing in some way—and that would be apparent to anyone who returned after a few days' absence. That person—let's call him Tom—would see the change that you do not. You, who had been watching the plant all that time, and Tom, who arrived after a few days, would both be looking at the same plant that you had been looking at continuously for all that time. And time will have passed. So just as we noted with Zeno's arrow, we cannot see change happening. We can see only the effects of change; in other words, after something has happened. Time is change, but change that has already happened: time is the effect of change.

Why is that? Why can we not see change as it is happening? One of the reasons lies in the backward-to-memory shifting *now*, which moves infinitesimally slowly—so much so that we do not notice it happening. It is our not noticing the imperceptibly slow movement of *now* back to our memory that gives us the perception of *the present*. If that is the case, why don't we notice a change when staring at the plant over a long period of time? The reason is that the mind 'gets used to' those tiny changes as they occur and are moved into memory. Putting it another way, as each change goes into memory, the change is so small and frequent that we can't compare it with the memory of the one before and therefore, we don't recognize that there has been a change. As far as we are concerned, nothing is happening and we are in *the present* ('now'). But when Tom comes in and looks at the plant, he notices the change that has occurred in his absence, comparing it to what is in his memory. Not having sat through the slow movement of change, he recognizes the difference. Something similar happens when you see someone you haven't come across for a time, while those who have seen the person during that period will not have noticed the changes occurring.

A dramatic, descriptive example is Miss Havisham in Charles Dickens' *Great Expectations,* in her white wedding dress and white bridal shoes, with a bouquet of flowers. When Pip came upon her first, he saw that what should have been white had turned yellow, and that the person herself was no longer young but an old woman. For Miss Havisham, though, there had been no change. She still saw herself as the young woman she had been in gleaming white. Pip never saw her in white; his initial view was of what he imagined she might have looked like before it all withered. Pip, of course, saw the change after it had happened.

It should be clear that whether the change is slow or fast, it is seen always after it has happened. Watch the sun setting. You can't see the sun actually moving, however hard you try, but you are aware that it has moved closer and closer to the horizon, until it disappears below. And so it is with a candle that melts as we watch it, seeing the state of the candle fractionally after each minimization.

And while we are at it, we have an explanation at last as to why a kettle seems to take much longer to boil when you watch than if you take your eyes off it. Think of the plant and there is the answer. Watching the water being heated is a recipe for time seeming to slow down, since you are comparing very small changes. Look away and the water will have boiled really quickly, the reason being that the change that occurred in the meantime is a jump from the earlier state in your memory. Similar is the reason time seems to take longer for the person waiting than for someone who is making their way to a meeting. And, of course, it accounts for the slow crawl of time when we are bored compared to the galloping of time when we are having fun: being aware of time passing is similar to watching water being heated or paint drying.

Already mentioned—and I do so again because it is so important—is that not observing change as it happens is responsible for the feeling of 'living the moment,' the so-called *present,* while we actually experience change only after it happened.

The past, then, is paramount in time. In the absence—the impossibility—of *the present,* it is the past that is the anchor of our existence as humans. The past goes on and it does so continuously, moving back further into memory.

For a person without memory, there is no past; what they cannot access does not exist for them. On a broader level, that which is not part of collective memory does not exist and is, therefore, not part of a narrative history. Put succinctly, anything that exists is part of memory. No memory = no past = no knowledge.

Before we leave memories, it's worth returning to a point we looked at previously: memories that slip back and simply disappear. If we have totally forgotten an event, what has happened to the event itself? Was there an event, in other words? The instinctive answer is that yes, there was, but that we don't remember it. But does that make sense? How can we refer to an event of which we have no knowledge? And what of an event that we don't remember but that someone else remembers? The fact that it is someone else's memory is no more than hearsay and does not make it part of our experience or of our past.

There are places I remember all my life
Though some have changed
Some forever, not for better
Some have gone and some remain

All these places have their moments
Of lovers and friends I still can recall
Some are dead and some are living
In my life I loved them all

And with all these friends and lovers
There is no one compares with you
And these mem'ries lose their meaning
When I think of love as something new

And I know I'll never lose affection
For people and things that went before
I know I'll often stop and think about them
In my life I loved you more

Paul McCartney; John Lennon, 1962

III

Will Be

Never there

So where are we? Certainly not where we thought we were. If at this point we're feeling somewhat unsteady, it's not surprising. After all, we have shaken some normal day-to-day, taken-for-granted perceptions. Not only have we discovered that *the present* may be more of a mirage than a 'fact,' but also that the importance of the past is tempered by the realization that we can't be all that sure of what actually happened and if, in fact, anything did happen.

But before we continue the questions and doubts about us within time, we can point to one certainty: the future is always ahead of us. Always. "In three words I can sum up everything I've learned about life: it goes on," said American poet Robert Frost. Whether optimistic or pessimistic, we cannot escape the future, which is always beckoning. This cannot be refuted since, to put it succinctly: when there is no future, we won't be alive to know. "I will, therefore I am," may be just as apt as Descartes' famous "I think, therefore I am."

Being aware that there is a future does not mean that we know what will happen in that future. There is no connection between knowing there is a future and knowing what that future will hold. The reason we never know what will happen in the future is that we never get there. Never. As sensible as it sounded, when computer pioneer Alan Kay said, "The best way to predict the future is to invent it," he was wrong. Even the prediction that we will be dead will never be seen by us to have been right—about others, yes; but not about ourselves.

Never here, always there, tomorrow is always ahead of us. There are no certain predictions about what tomorrow will hold. We are, though, pretty sure that what is here today will not have changed dramatically tomorrow. That's what we would hope. Although we are always on the verge of the future, of things about to happen, we nevertheless expect life to remain steadily familiar. When we are with somebody, we expect that person to look the same after a second, a minute, an hour... That is why we are so surprised to see an old photo of someone we know: we were expecting the person on the photo to be as we know them now, not as they were on that photo. And it is all so fleeting, for what we are anticipating is about to become the past.

Let's say that you have a dentist's appointment for tomorrow morning at ten o'clock. At 10.00 a.m. there you are at the surgery as arranged. Were you right about the future? Yes, you were (even though something could have happened to change the result). You haven't, though, arrived at the future. You could only see that you were right when the event occurred, no matter how certain you had been about it beforehand. But whether it turns out that you were right or wrong makes no difference, since whichever way, you know only afterward. We always get to what **was** tomorrow, it then being today. And today? When and where is it? Well, today is the tomorrow you worried about yesterday.

Our draw toward the future is relentless, so much so that the anticipation often surpasses the event itself when it occurs. When asked, a surprising number of people stated that they loved Friday, a working day, because it came just before the weekend. It's not, they said, that they preferred being at work, but that the anticipation was delicious. "The idea of the future, pregnant with an infinity of possibilities," said Henri Bergson, "is... more fruitful than the future itself, and this is why we find more charm in hope than in possession, in dreams than in reality."[96] Strangely—but not really—is that Sunday afternoon is already tinged with anticipation of the following day back at work. This is the other side of the coin to our discussion in the first chapter about being more fully able to appreciate events after they happened, rather than when they were actually taking place. What is it that keeps us from fully enjoying 'the moment,' but rather at a distance from it? It is the backward-moving *now* that is at fault again! As we have noted, we are never in *the moment*, never *now*, never in *the present*. The instantly moving *now* is already a memory when we catch what we feel is the present. What we have on Friday is anticipated memory of the following weekend, and also on Sunday of the following Monday. We are always in anticipation, as we will see again and again.

The iron-clad principle that there are no certain predictions about the future has not dampened legions of people making striking prophesies about what will definitely happen. 'The end is nigh'—a wishful-thinking warning—has been carried on signs around the streets for, well, it seems like forever. It is certainly taking its time, this nigh. The number of charismatic leaders—often linked to religion of one sort or another, who led their followers up mountains and into the wilderness for that final day, only to stare at the dawn of the following day and sheepishly explain the reasons for the prophesy not having been fulfilled—is large. Next time lucky, is the hope.

The notion of the end of time, though, is a basic creed of well over half the population of the Earth. Christ's Second Coming at the end of days as forecast in the New Testament's *Revelations* is mirrored in Islam's end of days, but with different main protagonists.

Surprising, then, is the fact that cutting-edge science also looks to the end of time. There is the Doomsday Clock, a symbolic clock face, maintained since 1947 by the board of directors of the Bulletin of the Atomic Scientists at the University of Chicago, Illinois. The closer the clock is to midnight, the closer the world is estimated to be to global disaster as a result of nuclear weapons, climate-changing technologies, emerging biotechnologies and cybertechnology. Moving toward the future and never reaching it, nobody will be around to witness doomsday until it is behind us. Doomsday is, and always will be, ahead of us, whether it is based on reasoning in science or on extraneous ideas of belief systems.

Never actually being there, the future is a make-believe place. As a place that is always ahead of us but that we can never get to, it enables our imaginations the freedom to soar. With all the mind games and mental gymnastics this allows us, the future-as-potential is a springboard for roving thoughts about time itself.

So while the past is alterable by our memories, the future is unalterable in the sense that we can alter only events that *lead up to* the future. There is an interesting difference here. Our past is altered unintentionally by us (so much so, that we usually aren't aware that it is not what we think happened), while possible future events are perceived by us to be open to intentional change.

Actually, belief in what we assume will happen is very useful; it is part and parcel of our thoughts. All our actions are based on what we expect will happen in the future. Without that assumption, we wouldn't do anything. Knowing that our plane will take off at 7.00 a.m. is what gets us out of bed at a certain time to enable us to catch the flight. We are very disconcerted when the future is changed; for example, if our plane is delayed or canceled.

"Procrastination is the art of keeping up with yesterday," said humorist Don Marquis. Perhaps that is a good way of summing up the future: a continual movement toward yesterday. Or, to paraphrase Paulo Coelho, "Putting things off till tomorrow is a good way of never getting anything done."

Time does not stand still and so neither do we. Locked within a circle of time that progresses through its own natural momentum, we constantly look forward toward the inevitable next point. Getting used to something new, taking it for granted and then craving for something to ameliorate it, is a natural human condition. Even as we achieve our dreams—finding true love, buying a house, making a million—our expectations rise at each stage, resulting in no permanent gain in happiness. Perhaps this explains why winning the lottery often turns out to be more of a curse than a blessing. The future is a demanding beast, always wanting more and never giving us the satisfaction of relaxing with what we have achieved.

The urge to know what will happen is so strong as to sometimes become a self-perpetuating belief. Not only clairvoyants feed on this belief and hope (and perhaps even believe in it themselves), but a whole discipline has arisen, involving experts in various fields. Known as 'Superforecasting' (based on a book of that title, by social scientist Philip Tetlock), it has spawned a veritable industry that includes 'master classes' and 'forecasting tournaments,' in which selected candidates were asked questions about the course of events. Some of the best minds in their fields take part in these projects. As Steven Pinker, who has written about Tetlock's work on Superforecasting, said, "Everyone agrees that the way to know if an idea is right is to see whether it accurately predicts the future." But we can't know that. We can know only if an idea *was* right. However scientifically and information-based the predictions are, we cannot know whether they panned out until afterward, when they were seen to have been correct or not. The fact that, as Tetlock shows in his book, some people are better than others in predicting the future, this is also only proven after the events have taken place. And it goes without saying that even those 'Superforecasters' are seen to have been wrong in many cases.

Yet, we keep trying within a human obsession to know what is in store for us, to rattle the cage of the future, of time itself. Of all the dimensions, it is time that we cannot do anything about. As a perpetually dangling bait, the future is frustratingly always ahead of us. Stories of time travel are a manifestation of this frustration and the concomitant belief that it is possible to get to the future 'if only.' Einstein's special law of relativity adds theoretical hope to that possibility, since it shows that time runs faster or slower depending on one's location in space. It is the reason physicist Brian Greene states categorically that "time travel is absolutely possible,"[97] and that "relativity lays out a blueprint for time-travel to the future."[98] *It doesn't.* We can't slow down or speed up the rate of our personal time. Clocks and watches do not go slower or faster for us, no matter where we are or how fast we travel.

If you could speed away to the stars at close to the speed of light and return after 3 years, although everyone on Earth would be 67 years older, you would still have aged only 3 years. And, yes, it might be nice to note that you had added on only 3 years while everyone else was 67 years older, the fact is that you had not really changed anything in your personal time span; you would still be only 3 years older than you were when you left. Unfortunately, despite the fact that everyone you knew on Earth would be 67 years older than when you had left 3 years previously, it would not mean that you would live longer than you would have had you stayed on Earth.

In fact, you wouldn't even think of it as having 'traveled in time' if there weren't anything to compare it to, if you weren't later in contact with those who had aged at a different rate; time, after all, is relative. Everyone's time is personal; there is no time in a universal sense. Clocks do not tick time at different rates; they do so only in relation to each other. In a colorful video demonstrating 'time travel,' particle physicist Brian Cox says that "moving clocks move slower."[99] They don't. Clocks do not slow down. Things change in relation to one another, so that time changes only in relation to the time of something else. It's why we each have our own past and our own probabilistic future. What Cox's demonstration shows is that "time is personal," as he points out, but *not* that the person who was doing the demonstrating "had travelled in time."

As we have seen, our getting 3 years older in outer space while everyone else's age increases by 67 years here on Earth has to do with our own personal time. It is not time travel. Imagine two friends, Mary and Carol, both 25 years old. Mary was the lucky astronaut who traveled into space for 3 years at close to the speed of light. When Mary returned to Earth she was 28 years old and found Carol to be an old woman of 92. That would be quite a shocking reunion, but possibly mitigated by the fact that Carol thought she could ask Mary—who, theoretically, had time-traveled—what was going to happen in the future, perhaps to know the winning lottery numbers that were still to come. Mary would look at her in amazement, for the lottery numbers that had come up were obviously not in Carol's future. Mary was by now thoughly confused, since it seemed to her that it was Carol, who had traveled in time; after all, Carol had lived through 67 years while Mary had been through a mere 3 years. And they would both be amazed to hear Brian Greene stating that "If you wanted to leapfrog into the future, to see what the world will be like a million years from now, it could theoretically be done." Mary would, in fact, have no idea what Carol's future will be, since she had not traveled to the future. Time had *not* gone slower for Mary. When she came back after 3 years, those 3 years were in her past, her memory.

The notion that there is such a journey into the future is patently ridiculous. Mary didn't travel forward in Carol's time or, indeed, in anyone's time. She didn't travel forward in her time any differently to what we all do as time passes. Neither of their clocks moved faster or slower. To illustrate how nonsensical the time-travel notion is, had Mary moved forward in time, we could likewise claim that she subsequently moved back in time when she returned to Earth, so changing the future that she had just been to! In fact, where she had been to was in the past for her, not in the future. It certainly wasn't Carol's future. By the same token, we could say (if we wanted to further mangle our minds) that for Carol to then do the same trip would be for her to travel to Mary's past.

If the mind game we just set out was confusing, that is not surprising. The fact that it is deemed to be theoretically possible is even more surprising. When it comes to traveling back in time, though, this is generally accepted as impossible, owing to the implications that make it so. The 'grandfather paradox': going back and murdering your grandfather before he's even born. If you've killed your grandfather, you've prevented your own existence, and if you never existed, how could you have traveled back in time in order to do it? If that means that backward time travel is impossible,[100] why would physicists allow forward time travel as theoretically possible? The implications would be just as impossible, as we have seen.

Beneath the concept of time travel is the implication that the future is where we have not yet arrived. This is, in fact, what Einstein's theory of special relativity suggests: by destroying any notion of universal time, and having time relative according to location in space, the future is supposedly already out there waiting for us, which we simply can't see until we get there. "Are we there yet, are we there yet, are we there yet?" asks Bart over and over in a car ride with his family in the television cartoon series *The Simpsons*. If the fact that we are moving linearly toward the future implies that we will eventually arrive, it has no connection to time travel as is popularly envisaged.

And underlying it all is the covert presumption that there is a future real or correct time that can be used as a marker to show that time moves faster or slower. But there is no universal time; time depends upon where one is. Time is always in relation to someone else's time, so that Mary's time is always relative to what Carol's time is. If, when Mary was traveling away from Earth, the Earth was destroyed, she could not be considered as traveling in time, since there would be no way of comparing. In any case, there could never be a way of comparing two times; it would not be possible to communicate with someone in another 'time zone,' because the length of time it took for the communication to travel between two time zones would equal the gap in time between them.

It's time, then, to put to rest the myths that have been insinuated into our imagination as to what it could mean to travel forward in time. Like Santa Claus, time travel to the future is an idea, pleasant in its promises. But, unlike Santa Claus, who we eventually grow out of, the romantic possibilities inherent in traveling to the future, from *The Time Machine* of H. G. Wells to TV's *Doctor Who*, have never been put out to pasture. The question is not whether it will ever be technically possible, but whether the notion 'travel to the future' itself makes sense, even in terms of science fiction.

Predicting

The belief that we frequently predict the future is so strong, that it can be thought of as a type of wishful thinking, where looking forward to what might be is often better than considering what is. Nothing is as good as what you haven't yet done. It is certainly safer and more comforting. As the future-directed counterpart of memory, we anticipate—sometimes accurately—future happenings (as is memory of what are considered by us to be accurate past happenings), and experience emotions in the future such as fear, dread and hope. This can exert a powerful influence on our backward-moving *now* (as are emotions in our memory that also influence our so-called *now*). In short: wishful thinking of the future is the reverse side of pleasant memories of the past that serve as escape refuge to nostalgia. This is nicely illustrated by Slavoj Žižek, who tells about a man who has a wife and a mistress, and wishes for something to happen to the wife so that he can be with the mistress permanently. But, says Žižek, "he would not be happy if that happened, since really he wanted the mistress to be at a distance, just out of reach, not completely attainable." The possibilities of the imagined future are more exciting than the actualities of the controlled past.

Just as we can't know what actually happened in the past, so we can't know what will happen in the future. The future is supposition based on past experiences, which themselves are uncertain. The further back our memories, the more uncertain they are. The same with the future: the further forward we try to see, the more unreliable it necessarily is.

Everything human is finite. The future and the past are both finite: for us, our personal past begins with birth and will end with death. "Hope springs eternal in the human breast," said Alexander Pope. Hope can lie only in the potential future, despite the knowledge of it having to end, and despite the fact that the future is always shortening and linked back to the increasing past. Sadly, hope for the future often becomes hope in a distant past, as is manifested in those suffering from dementia.

We tend to describe both the future and the past in plus/affirmative terms (what happened or will happen), rather than in minus/negative terms (what did not or will not happen). In other words, just as the past is about what we believe did happen, the future is about what we believe will (or might) happen. We cannot consider the past as referring to events that did *not* happen, nor the future to be about events we believe will *not* happen. We have 'if' to help us with those impossible scenarios.

Sometimes we feel as though we catch a glimpse of the future. If we read a newspaper article from, say, last week, we are in the interesting and somewhat smug position of already knowing if the supposition in the article turned out later to be correct. It's a strange feeling, knowing the future at the time we are reading the paper. But we don't, of course. At that point, it is the past we know about, even when we read an article that predicts the future we already know about.

Those news flashes that come to our smartphone or to our

computer screen: if instead of reading them in chronological order, we do the opposite—rather than up-down (from the latest to the earliest), we read them down-up (from the earliest to the latest)—we would get a different perspective of time and the world. A veritable move back to the future.

In Umberto Eco's book *Numero Zero,* Simei, publisher of the newspaper *Domani,* explains to his proposed editor what its aims are in the demo issues that will be published. In relation to a terrorist bomb that went off a few months before, he wants to show how the newspaper would have treated the story at the time. "In that particular case," he said, "we already know what will happen, but we'll be talking as though the reader doesn't yet know... In other words, we have to say to our owner: this is how *Domani* would have been had it appeared yesterday. Understood? And, if we wanted to, even if no one had actually thrown the bomb, we could easily do an issue *as if.*"

It is the 'if' that gives wiggle room within a make-believe scenario as potential in time. We could call *if* the non-committed state or fulfilled potential in all different times. So when we say, "If I see you after work, I'll give you a ride home," we are talking about a future event that will already have happened—in other words, a potential that has been fulfilled. Or how about: "If the sun disappeared, we'd be in trouble."? Also, in the following example, we have a past situation, in the future: "If you had arrived earlier, you would have seen the game." 'If,' in fact, may be the only way we can travel in time, using 'if' as our time machine. It's make-believe: the potential of the future that we can play around with, since we fulfill a potential that could never normally be fulfilled.

When referring to the future, grammar is very cool. It allows us to express the most complex thoughts about a time we can never reach. The non-reachability of the future is encapsulated within thoughts such as: "It's ten o'clock. By now Jill will have arrived home." It also allows us to discuss things that will happen in the future while referring to them in the past: "Will you have eaten when I arrive?"

In which way, though, is there a future? If it can be said to 'exist,' it is only as potential. And as potential, the future is not of what *will* happen, but of what can *possibly* happen, perhaps even *probably* happen. Even for events that we are absolutely positive will happen, we can go no further than probability (even if we consider it to be a 100 percent probability). A classic example is that of the sun, which we are sure will rise again tomorrow, as it has done since the Earth was formed. The point here is not a discussion as to whether or not it will rise tomorrow, but the fact that we will know if it has only after it has done so! A prediction, in fact, of what *will have* happened.

The 18th-century Scottish philosopher David Hume used the sun example to ask about our beliefs in the future and whether they are justified. There is no doubt in our minds that the sun will rise tomorrow, based on our experience that the sun has always risen. And, of course, we have scientific laws that show why it will happen. But why do we believe they will hold tomorrow? Very naturally, because of our experience that these laws have held in the past. But is this induction justified? It certainly feels so; but really it isn't, since we actually have no way of knowing that what happened in the past guarantees it will happen in the future, despite our rational induction that it will.

With the future always ahead of us, we will not know if anything is 'true' (that it will happen) before it has taken place. Surely, the future cannot be true today, no matter how accurate the information on which prediction is based. But what happens if our prediction is found to have been right? Does it only *then* become true (true that it happened) or was it true before (true that it will happen)? And if it *was* true beforehand, is it just that we *didn't know* it was true? The point is that we *can't* know today about what will be; we can only know after it has occurred.

Coming back to our dentist appointment, it is tempting to say that it will always be true that we were at the dentist on that day at that time—a state that British philosopher A. C. Grayling refers to as a 'timeless truth.' "Truth can become permanent," he said, pointing out that nothing could change the fact that the meeting he was attending was taking place. "It would always be true," he added, "that the discussion [we are having now] had taken place at this time and at this place, no matter what people would say about it in the future."[101] In other words, for us to believe that something is true, we necessarily need to think of it as being always true. But can anything be forever? Timeless, without changing? In the light of our foray into the vagaries of memory and the past in the previous chapter, in which way would Grayling's meeting be 'permanently true' in, say, a hundred years, if its contents were being reported differently over time? The 'truth' about the meeting would, surely, be different then, since there would be varying views about the event. But Grayling was referring to the fact that the meeting took place, to which the question must be: If there was no record of the meeting itself, where would it be for anyone to refer to when there was nothing to refer to? Surely, without any memory or reference to the event, it would not exist in any way, and so, for all intents and purposes, would never have existed.

Predictability is the bedrock of how we conduct ourselves according to what we believe is ahead of us. Essentially, predictability is repeatability, and as such is part and parcel of daily life, from computer programming to sports. If we throw a ball, and know its weight and speed, we can calculate and therefore predict exactly where it will land.

Does that mean we can predict the future in these cases? The answer, unfortunately for science, is no. The reason is that all of our predictions work only in theory—when the conditions are perfect, when nothing untoward happens. In other words, in a perfect (nonexistent) world. In our world, predictions are never in *perfect* mathematical agreement with the observations and experimental measurements. As mentioned early in this book, timing and, in fact, all measurements, depend upon the accuracy of our instruments. Things work because the results are approximate. When measurements are repeated (as they must be, preferably by independent observers), they scatter randomly around some average value. Prediction is confirmed when it lies within an acceptable error.

When science predicts the future within an acceptable error, it does so because 'acceptable errors' can be so forgiving as to ignore unforeseen circumstances. It is those circumstances that cannot be predicted! Science can predict what might happen only if all the conditions work out. People landed on the moon in 1969, but Apollo 13 failed to do so in 1970, when an oxygen tank exploded on the lunar lander. In 1986, with more advanced systems, computers, software and mechanics, the space shuttle *Challenger* broke apart 73 seconds into its flight, leading to the deaths of its seven crew members.

The future is simply inaccessible, despite technology considered to conform to the most brilliant precision and accuracy. Quantum physics, currently the most exactly accurate physical theory, produces measurements that can be thought of as almost infinitely accurate. Yet, results of quantum events can never be perfectly repeated, and so cannot be predicted, either.

Evolution is a prime example—a combination of *what was* with expectation of *what will be*. As a description of how life on Earth evolved through the change in heritable traits of biological organisms over successive generations due to natural selection and other mechanisms, the question is whether evolution can predict future processes. Generally accepted is that if the clock were set back and the whole process of evolution began again, it would happen the same way. Caribbean lizards are often given as an example, where the Anole lizards are shown to have evolved independently into similar-looking forms, despite living on four neighboring islands—Cuba, Hispaniola (the countries of Haiti and the Dominican Republic), Jamaica and Puerto Rico.[102] There are, though, contrary opinions as to the inevitability of evolution on Earth. Paleontologist and evolutionary biologist Stephen Jay Gould claims that evolution could have gone in different directions had there been any number of small events, such as a storm that swept away a particular pond, or, say, a poor season for insects.

Whether evolution would have happened exactly as it did, or whether it would have happened differently under disparate circumstances—or how humans, and indeed, all of nature, will turn out in the distant future—the results were unpredictable then and are unpredictable for how evolution will continue. Evolution was, and is, probabilistic. That doesn't make the theory of evolution any less valid; all it means is that we see only in retrospect its brilliance in explaining natural phenomena.

That brings us naturally to the giraffe, a seemingly straightforward example of evolution. Its long neck, forelegs, head and tongue, gives it an advantage over other animals, allowing it to eat from the highest branches of trees in order to survive in areas where there is a dearth of food lower down. Commented on by Darwin in the sixth edition of *Origin of Species* in 1872, and already some sixty years earlier by French biologist Jean-Baptist Lamarck,[103] and by others later with differing views, it seems obvious that the giraffe's unique characteristics cannot be predicted except in hindsight. Why didn't other animals acquire its advantage? There is no conclusive answer to that question and no definitive way of predicting what nature will yet have in store.

The principle of non-predictability goes as well for premonition—the strongly felt claim that we knew in advance what was going to happen—the proof being that it did happen. And there is the catch: it **did** happen. Now while prediction more generally follows inductive reasoning based on what has happened in the past, premonition, with its mystical or magical aura that gives it its strong feeling, is little different from any other look into the future. It is seen afterward to have happened or not to have happened. Not premonition, but rather postmonition. To put it crudely, the proof of the pudding is in the eating: it will be tasty (or not) when it is eaten, not before.

It seems we can't escape the past, not even in the future, since they are inexorably bound to each other. Disconcertingly, we find that our strong ideas about what will happen can be based only on what we believe will be shown when they are in the past.

In that spirit, the black swan theory, developed by Nassim Nicholas Taleb, describes the phenomenon of events that were completely unpredictable and therefore come as a surprise because they are so rare (hence, 'a black swan'). Those events are beyond what could be expected, have a major effect, and are then often inappropriately rationalized with the benefit of hindsight. Examples are 9/11, the collapse of the Soviet Union and the ubiquitous spread of the Internet.

An excuse often made when failing to predict an event is that it couldn't have been predicted because it had never previously happened. As an excuse it is pretty lame, especially when it is given by those whose business it is to make predictions in their particular field. When there is no relevant past to lean on, predictions are mute. Taleb cites Alan Greenspan, the former chairman of the US Reserve Bank, who made that very point when attempting to explain to Congress why he had not seen the banking crisis of 2008 coming. And there is the case of Microsoft chief executive Steve Ballmer's interview in 2007, in which he said, "There's no chance that the iPhone is going to get any significant market share." The rest is history: seven months later, Apple's first smartphone went on sale and redefined the mobile phone industry, enabling millions of people to have computer power in their pockets for the first time.

The notorious difficulty of publishers to predict the success of books is because they naturally look for something on the lines of what was successful previously. Every so often, there are surprisingly unexpected hits (*The Bridges of Madison County*—a short romance—comes immediately to mind)[104] that break away from the current mold, changing it and unwittingly being the first of a new series of clones.

And, of course, there are the spectacular flops onstage and screen following previous great successes by the same motivators. Incredible it is to see how the late producer Richard Zanuck made flops like *Hello Dolly!* and *Doctor Dolittle* right on the heels of a blockbuster like *The Sound of Music*. Not within Zanuck's vision were the shifting currents that had gone from the golden era of the early 60s to the chaos of Vietnam and the Kennedy assassination. That audiences no longer wanted light-hearted musicals was not noticed within the aura of what worked in the past. But then we can all be wise in retrospect.

And what about flying cars, which all children growing up in the mid to late 20th century assumed would exist by now? Egged on by magazines, such as *Popular Mechanics*, they had justified adult reasons for their predictions. Where are the forcefields, teleportation pods, antigravity skateboards, immortality drugs, colonies on Mars and all the other technological wonders that seemed sure to be around in the future? Based on advances in science, it did not seem unlikely that a trip to the moon would be as easy as going for a week to a Greek island. Much of what was science fiction at the time is still in the future, even though we are closer to some of it being realized. But the future was always almost there, wasn't it? Almost, but not quite. In 1989, the movie *Back to the Future II* had flying cars and antigravity hoverboards used by ordinary teenagers in the future, the year 2015. But when, 26 years later, we eventually got there, 2015 evaporated into the known past, turning out differently from when it had still been the future.

How can it not, in fact? Some of the predictions came to fruition, while in 2015 others hadn't yet but were within sight. All of them, though, were based on human imagination, on ideas that work in some form. There are no predictions, however wild, that are not based on something that had never been thought of or imagined. The future, then, is a prediction of the past! It is anticipated memory. When Alice complains, in *Through the Looking Glass*, that she cannot remember things before they happen, the Queen responds testily that "it's a poor sort of memory that only works backwards."[105] While Alice claims that we have no memories of the future, this is not quite accurate, as we saw with déjà vu. "Sometimes I have these premonitions and I don't forget them, so I will be prepared when they happen," says Isabel Allende in *Island Beneath the Sea*. And we do have imagined memories of a future based on scenarios fed by past memories. The future cannot be anything other than anticipation based on whatever has already happened or imagined to have happened.

The historian Sir Lewis Namier puts it succinctly, in opposition to Alice's logic, when he states that by using our memories as stepping stones to the future, we eventually remember what has become the future. "It is to be expected," he says, "that people remember their past and imagine their future. But in fact, when they write discourses about history they imagine it through the prism of their own experiences and when they try to ponder over the future they refer to presupposed analogies with the past, until, in a double process of repetition they imagine their past and remember their future."

The link between past and future is so strong, that the loss of one (memory) impairs the ability to produce the other (prediction of the future). Memory loss has an effect on being creative. This should be obvious, since we can think about that which is based only on what we have in our memory. In extreme cases of memory loss, it was found that the ability to plan for the future, and even the capability of thinking about the future, was lost, as well. It isn't surprising, then, that there is a link between memory and genius. Renowned geniuses had prodigious memories. The French mathematician Henri Poincaré—perhaps one of the most creative minds of his, and perhaps any, time—had a memory that was nothing short of remarkable. Able to remember the exact page and line of a book where any particular point had been made, he could also retain information that he had heard, absorbing lectures without taking notes.[106] And let's not allow ourselves to be fooled by Einstein's so-called day-to-day forgetfulness, which had nothing to do with his incredible memory when it came to the imagination involved in the genius of his physics work.

The past and the future are so bound up, that sometimes the resultant stew becomes amusingly playful. Light comes to us only from the past, showing us what has been. Looking up at the sky, we see stars as they had been thousands of years ago, the amount of time it took for light to reach us. But if we were on a rocket going into space, the stars are in the future to where we are heading, with the past ahead of us.

Back on Earth, the moment we plan what we are going to do, that itself becomes memory; so it is on that memory that we base our movement toward the future, one of those cycles that moves us along our path from our beginning to our end. Since the future as potential is always ahead of us, the movement is continuous and unrelenting. Look at the process by which we plan a journey or a holiday: such a trip requires the creation of future events that become the past as soon as we visit what we had planned. Simply put: when we plan the future, we are planning what will be the past. Nietzsche's point that "The future influences the present just as much as the past," would have been more accurate had he simply said that the future influences the past.

Without the past there is no future; without the future, there is nothing.

It should not be surprising that our future is unknown and

unknowable: if, as we have seen, the past can be accessed only through memory, which is unreliable to say the least, any future (all of which is based on memory of the past) is built on a house of cards. We do know that the past occurred—we were, after all, younger than we are now—and we assume that the future will occur because we will be older then. And just as we cannot be sure about what actually happened to us in the past, similarly we are unable to access the future that leans on predictions based on that past. In a nutshell: both the past and the future are inaccessible to us, even though we know that they 'exist' in some way.

It gives us a reasonable explanation as to why advice is so nebulous, built as it is on past experience, refurbished and recycled as fresh and useful.

The symbiotic relationship between the future and the past has its psychological advantages. In many ways, the future is a safe refuge, not only comforting, but also essential to us. "We tolerate the present, but what we love is the future, which is about as far away from the past as it is possible to be," says one of the characters in *The Hired Man*.[107] That is what it feels like but, in fact, the future and past are inexorably connected in a loop. In Thomas Mann's *The Magic Mountain*, Hans Castorp reflects on Christmas that had just passed and that while lying in the past, it just as equally lay in the distant future, a year away, so that "twelve months would bring it round again."

Comforting and vividly real is the thought that however bad something is, however painful and terrible, there is the thought—the absolute knowledge, as far as we are concerned—that "this will end and, at some point" i.e. we will be able to look back on this one day. Because there will always be a future, there will never be a time when we can't look back at what was happening, even if the memory of it will have changed. Clear in all this is that the present becomes the past, always (and immediately), and cannot allow a static state of *now*. Luckily for us, there is no perpetual 'now'; if there were, we could land up in perpetual hell.

So there it is, the future that is always there and always on the verge of being reached. It's a vicariously good way of summing up the age-old question of 'What is it all about?': life consists of periods of waiting. Put succinctly, we are waiting for what is ahead to become the past. Obviously not everything 'in the future' pans out, so expecting the future to be real and definite is a recipe for the frustration of one unmet expectation after another.

As Jerry Seinfeld said: we always want to be somewhere else. When we're at work, we want to be home. When we are home, we want to be out. When we are out, we want to return home. It's not even that we want what we have once we've got it. Michael Foley calls this "the glamour of potential," a relentless churning of desire by which the things we have are devalued by the things we want next.[108] It makes no difference if things are going well or not. So when US philosopher Martha Nussbaum said, "If you have a good life, you typically always feel that there's something that you want to do next," it could just as easily be applied to someone having a terrible life, who wants to move forward to perhaps something better or simply to get out of the current situation.

In the previous chapter we mentioned the draw of nostalgia, where people are not satisfied with the present and generally consider the past to have been better. In that sense, we want the past to be our future. The seeming urge to go back is actually the desire to move the past into our future. It's on the lines of the impotent urge to be young but with the experience of being old. So here we have another cycle that drives us linearly forward: the urge to bring the past to the future while having the future become the past.

The looped intertwining of the past and the future says a lot about us, trapped as we are within what we were. And it's an explanation as to why we can never arrive at the future: the inexorable pull toward memory of the past. It is not, as humanist and law professor Roberto Mangabeira Unger puts it, that "anxious for the future, we lose life in the only time that we have, the present."[109] We stride toward the future on a metaphorical escalator that is itself moving in the opposite direction, back into the past into our memory.

We are continuously and involuntarily being sucked backward, away from the future that we will not reach. Like the Bible story of Lot's wife—who, despite God's warning not do so, did look back as she was escaping from Sodom—we also cannot avoid the magnetic hold of the past. So it is also in the mythical story of Orpheus and Eurydice, in which Orpheus is allowed by the gods to take back his great love, Eurydice, from the underworld of death on condition that he not look back at her during the journey out. But just near the end, when Orpheus no longer hears Eurydice's footsteps behind him, he looks back. Uttering a small shriek, Eurydice disappears back into the land of the dead.

In the same light, but less dramatically, is Mick Jagger, who, notoriously evasive in interviews, insists he never dwells on the past, claiming: "I don't want to erase it or to regret it, but I don't want to be its prisoner either."

But prisoners we are. We cannot help looking back, just as we continue looking forward. As much as the future is the guardian of the past, the past is the guardian of the future. Their interdependent cyclical existence combines the relevance of the past with the potential of progress in the future. Or, as the communications theorist Marshall McLuhan put it: "We look at the present through a rear-view mirror," so that "we march backwards into the future." Danish philosopher Søren Kierkegaard summed it up similarly 200 years ago: "Life can only be understood backwards but must be lived forwards." Mick Jagger having a child at the age of 73 is perhaps a personification of that principle.

The fact that the future is potential for what *might* happen, rather than prediction as to what *will* happen, has its advantages. How restricting it would be if we knew definitely all future occurrences. Instead, our imaginations can soar into possible futures, never knowing but always wondering.

What, though, if one day we were able to know what the future holds? Neuroscience seems to be moving in that direction, with ever-greater knowledge about the physical workings of the brain and the connections that signal between neurons leading to actions and even thoughts. Predictions as to individual human behavior are becoming neurological 'facts,' interpreted through the physical workings of the brain, in neurocriminology, an offshoot of neuroscience. The field is an ideological hot potato, though, because linking criminal behavior with genes and the physical characteristics of the brain seems suspiciously like an offshoot of the now-debunked 19th-century phrenology, where bumps on the head were analyzed to disclose behavioral characteristics, often used to support crude racial and class-based stereotypes of criminal 'types.' This continued with Nazi Germany's interest in eugenics, the aim of producing better human beings, perhaps the most taboo of all academic disciplines.

Although it is distinctly not politically correct to link behavior to nature rather than to nurture, tests made on criminals and those in death row in the USA purported to show physical differences in their brain compared to non-criminals. As controversial and open to interpretation as the tests and results are (and whether prisoners' brains and behavior had changed as a result of being in prison), the big question is whether it would be ethical to prejudge people according to brain scans and hence incarcerate them because of their supposed propensity toward crime. In other words, if we could 'spot a potential murderer'—which would mean actually predicting the future—should we place them under lock and key before they had committed a crime? And what about screening at school, so that students with the 'wrong' genes could be ejected, or perhaps treated ('cured') in some way. While it was still science fiction,[110] it was shown shockingly in the 2011 movie *We Need to Talk About Kevin*, in which the mother of a murderer was wracked with thoughts about what she could have done at the time she saw how he was growing up. A similar theme, illustrating the moral and legal issues, was played out in *Minority Report* in 2002, starring Tom Cruise, about a specialized police department that apprehended criminals based on foreknowledge provided by three psychics.

Before we continue down that road, perhaps we should consider whether it's worth getting hot under the collar about issues that might not be worth considering in the light of our foray into time. After all, as we have seen, predictions as to the future cannot be more than that, since we never reach the future. And since time is change, assuring that nothing is static, the discussion about the primacy of nature (genes and the physical workings of the brain) over nurture (outside influences) is sterile. How could we sensibly consider an irredeemable quality of humans, marked for life and untouched by time? The constant change that is time precludes even a momentary static situation. As we have seen, the real possibility of changing the past through altered memories, as well as the dynamic potential of the future, have within them constantly changing consequences.

But would those strictures apply to Artificial Intelligence (AI) as part and parcel of the evocatively named post-humans? The aim of AI is not only to simulate human intellect, but possibly to match and surpass it. While not yet doable, it is plausible; the way the field is progressing, it may not be unlikely that the future could see computers not only do more calculations faster and retain (remember) more than humans, but also be more intelligent and actually be self-aware. Being self-aware would mean that AI could set its own goals, strive to achieve them, and fight for them, possibly contradicting human interests. Imagine a scenario where a system decides to eradicate the human race by creating and distributing a deadly virus, as acted out in movies, including *12 Monkeys*[111] and a subsequent TV series of the same name.

All of this is in the realm of science fiction and therefore within our imagination. It's all part of a big 'if' as part of a possible future. But what it does is to take us to the edge of our imagination as humans and cause us to stand on tiptoes in an effort to look at the other side. We can't, of course. But when entrepreneur Elon Musk referred to AI as "our biggest existential threat," and warned that "with artificial intelligence we are summoning the demon," or when Stephen Hawking said that "the development of full AI could spell the end of the human race," and that "humans couldn't compete and would be superseded," we wonder if we shouldn't stretch a little further within our own discussions about time.

After all, if time is everything and everything is time, the question that needs asking is whether it is possible that anything can be outside time, in a time dimension different from that envisioned by humans: time that is not bound by human imagination. If AI can surpass humans and so is not bound by its subjective self, could it be feasible for AI to predict the future accurately, i.e. that everything it predicted would always come to pass?

At that point, AI would have become God (without the ethical and life-death characteristics humans have applied to God) and would be able to predict our future, even though we cannot. Moreover, unlike humans, AI would not know that there is an end, with no feeling of mortality.

Our musings into a possible apocalyptic AI future should

probably stop there. We can go only as far as our imagination lets us. If humans cannot succeed in seeing the future, we can but speculate as to how AI could. The notion that the human mind is just a massive supercomputer, and that future computers will therefore beat humans in every field, may not be the case. When it comes to solving problems that require more than the ability to crunch larger numbers more quickly, the human brain seems to possess an 'intelligence' that computers do not have. Ultimately, it is how humans and computers relate to, and work within, time. A computer dealing with numbers is actually a method of doing the job as quickly as possible. For humans, time is not always the be-all and end-all, so that they may ponder a situation and take a slower route to achieving a goal.[112] Time is still on our side, it seems.

Finite and infinite

It's another one of those internal paradoxes of time, the fact that even while there is always the future ahead of us, we know that time for us is limited. It's a strange phenomenon of human behavior that we know we are going to die, yet continue to make plans as if we are immortal.

Only humans realize their own mortality and thus are aware of time. In that sense, humans have a relationship to time; animals do not. Time and self-awareness are part and parcel of humanness, since awareness of time is awareness that there will be an end. As blessed as we are with the power of imagination, we are at the same time cursed, knowing not only that we will die, but also that we must. Yet, it is that knowledge, as devastatingly depressing as it is, that gives meaning to our lives; there is nothing more precious than knowing its temporality. In that way, the future represents purpose—the purpose to live. "Man has created death," wrote the poet W.B Yeats. Other creatures blindly struggle on, knowing only life until their moment comes. "Except for man, all creatures are immortal, for they are ignorant of death," wrote the Argentinean author Jorge Luis Borges.

In an important way, Jorge Luis Borges is wrong, though. While it is true that we are well aware of death, nevertheless for ourselves we are immortal, since there will always be tomorrow. Things will always happen as far as we are concerned, however far in the future they are. As Lord Alfred Tennyson put it in his poem *The Brook*: "For men may come and men may go, but I go on forever." Or as a line in Tarkovsky's movie *Solaris* has it: "We never know when we're going to die and because of that we are, at any given moment, immortal."

> Charlie Brown: "Some day we will all die, Snoopy."
> Snoopy: "True, but on all other days we will not."
>
> (*Peanuts* by Charles M. Shulz)

Goethe puts it rather more somberly: "Everyone carries the proof of his own immortality."[113] He also said, "Man should believe in immortality; he has a right to this belief; it corresponds with the wants of his nature."[114] And so many of us make New Year's resolutions, whether they are kept to or not (and most often are broken quickly, anyway), as a way of promising ourselves a continuing future.

A somewhat tongue-in-cheek view is given by philosopher Mark Kingwell: "The two-step law of life. Rule No. 1 is tomorrow we die; and Rule No. 2 is nobody, not even the most helpful robot, can change Rule No. 1."[115] In the light of our foray into time's vicissitudes, we can extend it to a four-step law of life. Rule No. 1 is tomorrow we die; Rule No. 2 is nobody can change Rule No. 1; Rule No. 3 is tomorrow is always ahead of us, therefore Rule No. 4 is that as far as we are concerned, we are immortal.

So does it end or doesn't it? Is it finite or infinite? It is—dare I say it—both. For us, as individuals, time is infinite. Because tomorrow is always ahead of us, we feel ourselves to be immortal within a seemingly infinite time span: when tomorrow doesn't come (when we are dead) we won't know. Possibly as part of an evolutionary trait, we need to feel this limitlessness of time in order to find a reason to continue. The supreme will to survive, present in most people, is the will to continue forever. It is the signature of all nations, peoples and tribes. All technological progress—in fact, progress of any kind—has as its aim to extend life and to do so into the unforeseeable future. As Dostoevsky, in *The Brothers Karamazov*, put it: "If you were to destroy in mankind the belief in immortality, not only love but every living force maintaining the life of the world would at once be dried up."

Here, then, is the paradox that is the tension within us: the movement toward what we know will eventually be the end, while also feeling the power of our immortality. As the author Max Porter says it so well: "We go to our deaths asymptotically, never getting there because 'we' and 'there' can't exist at the same moment, even in the case of those who have chosen to die."[116]

And so humans look for eternity within a limitless time—not necessarily in heaven (as do the monotheistic religions) but here on Earth. We always have. Searching for the elixir of life has been an obsession. From the First Emperor of China Qin Shi Huang (259 BC–210 BC) who searched for it, to the 21st century and the notion that modern science will create so-called transhumans, who will live forever though the replacement of organs, much like a car visiting the garage for new parts. And for the very rich, cryonicists aim to avoid the ancient resurrection problems by having their bodies frozen at minus 276 degrees Fahrenheit in liquid nitrogen, the idea being that future technologies will be able to revive them.

Failing that, and while we are still here, it seems that there is nothing better than the thought of death to focus us. Merely reminding people that they are mortal, that their time is finite, works wonders. A recent study on a group of basketball players found they played better if they thought about mortality before taking to the court. Even receiving instructions from a researcher who was wearing a T-shirt with a white skull on, along with other visual signifiers of mortality, improved play by approximately 30 percent in a one-minute basket-shooting challenge, compared to players who had not been shown those symbols.[117] It has to do with self-esteem, according to Uri Lifshin of the University of Arizona, Tucson, who puts it down to one's subconscious that "tries to find ways to defeat death, to make death not a problem... Self-esteem gives you a feeling that you're part of something bigger, that you have a chance for immortality, that you have meaning, that you're not just a sack of meat."[118] Not just sportspeople, of course. The researchers reckon that thoughts of one's demise will result in better results in a wide range of performance-based professions. The notion that our time is finite is, indeed, a sobering one. Against that is the equally strong feeling of continuity within an infinite stretch of one's personal being.

With the past and the future intertwined, there is no doubt that looking back gives fresh incentive to looking forward. It is the future, the period that we aspire to but will never reach, that is the most relevant to our conscious existence. Not the finite past with its possibly true memories, but the infinite future that is the proof of our being alive.

Lines in the sand

There are no lines. Time is change, and change is continuous. Simply because time passes, no line stays static. It's what was apparent in the extreme when we looked for *now*: where and when could *now* begin or end? That applies as well to the past and to the future, which are parts of time and do have duration. There is no point, though, at which the past begins or ends, and, as we have seen, there is never a point when the future begins (or ends). So while we can talk about events happening before and after, we are incapable of showing where the lines are. Interestingly, the most difficult tense for learners of English as a second language is the present perfect: it is not strictly past (although it would be translated as such into other languages) but a period that is a continuation of the past that connects into the present. 'I have been on holiday for a week' (present perfect) means that I am still on holiday, compared to, 'I was on holiday for a week' (past). The past seems clearer; after all, we know when we are no longer on holiday. But do we know? When exactly did our holiday end? Where was that line? Was it when we left our hotel? When we arrived at the airport or station? When we arrived home?

Time is a continuum. Because time, which represents change and hence an integral part of nature, is a continuum—everything is part of a continuum. It cannot be any other way, since everything is in time and time is in everything. And no matter how you define time, change is inherent within the universe. Yet, we are prone to draw artificial lines within continuums, a phenomenon that Richard Dawkins calls 'essentialism.'[119] At what point do we say that life begins? At the moment of conception? When exactly during development does an embryo become a 'person'? It seems so clear for some, as if there is a light switched on. Yet, an embryo develops gradually from a fertilized egg to newborn baby, so there's no one instant when 'life' could be determined to have arrived. It is we, according to our own criteria, who decide—which is why there are such violent arguments about abortion, with pro-lifers and pro-choicers willing to spill each other's blood over this issue.

Humans seem to have a problem with continuums and with the shocking notion that things may not be clear-cut, separated from others. This would seem to be the reason for the widespread feeling of there being a universal delineated 'now' that encompasses all of human time. Or the point in time when a prospective future event becomes a past event. There doesn't have to be such a moment, though, any more than there has to be a day when a middle-aged person becomes old, when day becomes night, when orange in a rainbow becomes red or when truth becomes a lie. What is flattery if not an indistinguishable line between the truth and lying?

It is the small changes, those that are on their own negligible, that can accumulatively make the big changes. An election is such an example. Why does one bother to vote, when it is, after all, just one vote? Or what difference can a grain of sand make? Yet, what is a pile of sand if not an accumulation of many grains of sand? The question is not how many grains of sand make up a pile, nor how many votes it takes to make a difference. Nor, indeed, how many drops of water make up a lake. We don't know the answer to those questions (even though we know what a pile is and what a pond is when we see them). Nor do we know how many grains we need to take off a pile of sand in order for it to cease being a pile. How much water does one take out of a lake for it to become a pond? Obviously, though, there is a change in perception when those actions are taken. These are examples of what is called the sorites paradox, which is thought of as being problematic due to the inherent vagueness of certain words, e.g. pile. So despite knowing what a pile of sand looks like, we have a problem with knowing at when it ceases being a pile.

The point, though, is not the vagueness of certain words; all words are vague, often ambiguous and change over time. It is that *everything* is vague in the face of constant and relentless change. As Johann Gottlieb Fichte put it in *The Vocation of Man* (1800), "You could not remove a single grain of sand from its place without thereby... changing something throughout all parts of the immeasurable whole."

What is vagueness if not an example of the continuum that is an integral part of nature? Again we come back to knowledge, which, however much we endeavor to characterize it, is itself part of a past-future continuum of time. So although we know what a pile is when we see it, we cannot adequately describe our knowledge of what that pile is.

But what about the vote? Unlike a pile, an election is not considered to have vague or fuzzy boundaries. We don't know how many grains of sand make a pile, but we do know how many votes a candidate needed to win an election—but only after the election has taken place.[120] Limited as we are about predicting the future, some people still play with the notion—indeed, are complacent about it—that 'one vote doesn't count.' Unlike the withdrawal of grains of sand one at a time from a pile, though, we can see how many votes we would need to subtract in order for the results to have been different. But since we cannot time travel, we are left with the results. Beforehand, elections are not less fuzzy than piles of sand. We could, though, make it easier by setting limitations: we could decide that a pile consists of 10,000 grains, so that any less would not constitute a pile. With elections, we decide that the winner needs just one vote more than their opponent (or opponents, depending on the system), the results being determined afterward.

And that brings us to time, the continuum par excellence. The sorites paradox does not apply to time; to think of time as *consisting of* moments ('nows') would not be accurate. We cannot reduce or add 'nows.' And with time not being absolute, time applies subjectively to each individual, making the backward-moving *now* strictly personal.

Any line, however straight it appears, will have a fuzzy edge for its border the closer you examine it. But wait! Surely a case can be made for there being some fundamental physical limits in the universe, a clear example being the speed of light (186,000 miles per second). That does seem to be an unbreakable boundary. But is it? What do we mean by stating that an object cannot exceed (in fact, reach) the speed of light?

According to Einstein's special theory of relativity, the faster an object travels, the greater its mass, so that when an object approaches the speed of light, its mass becomes infinite, as does the energy required to move it, and time stops. But then we are no longer talking about an object that is going incredibly fast, but rather something that has ceased being an object. What we have is a theoretical entity beyond that which is possible. We have a similar problem when it comes to the theoretical lowest possible temperature—absolute zero, which has a number: minus 273 degrees Celsius; but at that point the oscillations of molecules would become as slow as they could possibly be—in other words, they stop moving—so that anything at that theoretical temperature is no longer an object. To claim, then, that there are limits within nature is misleading; any theoretical limit leads to a change along a continuum inherent within nature itself.

The lines we ourselves draw are in all cases artificial, often decided arbitrarily by an authority. Thus, there are people 'above' or 'below' the poverty line. And being arbitrary, the line changes now and again. Death used to be officially designated as when the heart stops, but this has been changed to cessation of brain activity. And—casting our minds back to the beginning of this book—we cannot accurately notice even when the machine in a hospital started its long piiing, that denotes the end.

We are, though, constantly drawing lines. Lines that we need to get to, to hold onto, to cross. There is the best, the most, the greatest, the fastest, the cleverest. Or the worst, the slowest, the most terrifying, the stupidest. At what point does anything become the ultimate? In any case, when that happens—say the fastest runner, the fastest plane, the highest pole-vault jumper—it is by definition temporary.

Athletes are faster than ever, with records constantly being broken. Could there ever be a limit? Is there a physical limit for humans that cannot be crossed? Since 1968, when Jim Hines completed the 100-meter sprint in 9.9 seconds, being the first person to come in under 10 seconds, it has been done again and again, getting to 9.58 seconds by Usain Bolt in 2009. When the 4-minute mile was beaten in 1954 by Roger Bannister (3:59.4), it was a sensation, and in 1999, the record was down to 3:43.13.

Just as we wonder about the limits of human physical endeavor, so can we ask about the limits of mental effort and knowledge. As is often said, knowledge is the only resource that increases when used. But could human knowledge expand without limitation? With time as a process of change within a continuum, we can assume that there would not be such a limitation: time and space would continually expand as a way of accommodating the potential of knowledge. On the other hand, we did see earlier that knowledge changes over time. Perhaps, then, knowledge doesn't expand but simply changes. And let us not confuse knowledge and information, which is expanding exponentially with the increase in technological capacity. In that sense, the possibility of accessing ever-wider knowledge as information does seem to be limitless.

We are in a bind, though. If we don't draw lines to show limits, how could there be beginnings and endings? How does anything stop and start?

So when your car (or motorbike or bicycle or whatever) begins moving from being at rest, there seems to have been a moment before that and also a moment at which the car moved. That would indicate a clear beginning. Yet, in the light of what we are rediscovering about time, we cannot help but be faced with a conundrum. With time as an indivisible continuum, there aren't any moments. So how does the vehicle—in fact, anything—start?

There is an explanation. It lies in the instantly backward-moving *now*, which, as we have seen, has served as an efficient way of explaining many seeming-imponderables. A car that starts moving has, in fact, never been stationary. No object is still, but moves relative to everything else. For us, objects that we observe as in *the present* are constantly changing, albeit imperceptibly, through our *now* moving back to become memory. This is consistent with the continuum of time; there is no moment before the car moves and nor is there a moment after within a continuously moving *now*.

Let's take time itself: When did it begin? Purportedly at the Big Bang, when the universe and time flashed into existence. There is, though, a problem with the whole concept of time having a beginning. According to the Big Bang theory, the universe and time began at that point, so that there was no moment before the beginning of time.[121] However, it didn't begin from absolute nothing, according to quantum theories of physics, but from instantly appearing and disappearing particles, which are somethings. Calling it nothing, as does Lawrence Krauss in his book *A Universe from Nothing* even while admitting that space is not really nothing, does not ameliorate the position. It is not a matter of *when* time began, but *how* it did if there was no moment before the first moment. And the notion that there could be a first moment of a continuum in which there are no moments is far-fetched. Leaving aside the fact that we cannot imagine a moment that has no past, a backward-moving *now* does not apply in this case: there was no *now*, for there was no time into which it could move.

With the inevitable and constant change wrought by time, it is paradoxical that time itself could also limit us. Time as the limitless liberator, and time as the ultimate jailer. With time being everywhere and applying to everything, it does limit our freedom: we can, surely, never be free if we are never free of time. So while we have the urge to be free, and believe ourselves to be so, we are bound to time, because we are part of time and time is part of us. Ultimately, time is the boundary between being and not being—between being alive and being dead. For each of us, time begins when we are born and ends when we die. But, *as far as we are concerned*, that is theoretical only, since we are not aware of either event when they happen.

The very nature of a continuum precludes our discerning dividing lines between events or situations, even when we know a change is occurring that we cannot see. Where is the exact point where love becomes hate, light becomes dark, being awake becomes being asleep?

We're not very good at pinpointing causes, the reason being that there is always a preceding cause. Causes, like everything else, it seems, are on a continuum and are part of a spectrum. We could reasonably give the artist as the cause of a painting. But the material cause is the paint. And before that is the canvas, both of which are needed before the painting can exist. The final cause is the purpose for which it is made—its beauty or its message, let's say. Really, though, we have not completed our causes. An even earlier cause would be the reason the artist created their work. And prior to that, the scene they saw or the idea in their head. We could go back to speculating about the artist's upbringing or their genetic disposition. "Nothing remains true forever. In my experience, every book you write changes the conditions in which you write the next," says the author Howard Jacobson.[122] The notion that there is one unambiguous incontrovertible cause is suspect.

No beginning and no end. There is no final effect, just as there isn't a first cause. So while causes lead linearly in time toward the effect--in our example, the painting—we can see that it is not final, at all. Effects have their own effects, which are causes for other causes and effects. The effect of a painting is on the people who see it (not to mention the effect on the artist), and a chain of later causes and effects. Causality, we can see, is not exclusively a forward movement of events, but works circularly along that linear movement. It is the continuously backward-moving *now* that is instrumental within cause for effect and effect for cause. In that sense, the future affects the past as much as the past affects the future. *Now* is what causes change that causes change, within the universe that is perpetual change.[123]

Causes and effects are within the very fabric of existence. Together, they wend their way through from the largest to the smallest, with the butterfly effect a graphic example of the phenomenon. The notion that something as tiny as the wings of a butterfly in Brazil could eventually set off a tornado in Texas, is an exotic way of explaining why, for example, weather forecasting is often so wrong. All it takes is a small, unexpected occurrence somewhere to put paid to computational forecasts so finely calibrated. And it exemplifies again why future events are never reached, but seen only when they are in the past.

In the scheme of things, there is no difference between causes and effects, in the sense that causes can become effects and vice versa. As we have seen with a number of previous cases, the linear forward movement of time is bound up within a circular movement to keep it going. Put simply: things happen only because other things had happened before. Time is the engine of cause and effect. As pointed out earlier, the backward-moving *now* ensures that no object stays as it was, so that what happened is the cause of what happens later: a rock is there because it was there earlier.

Even with clear-cut beginnings and endings of musical pieces, however disciplined the orchestra, clear acoustic measurement devices will discern minute differences in the endings of all the instruments. Some avant-garde atonal music compositions that seemingly begin and end arbitrarily do that deliberately, so that the audience knows when the piece has finished only when the musicians pack up their instruments and walk out.

Without clear beginnings and endings, it shouldn't be surprising that progress in general is hard to fathom. For good or for bad, the very nature of mankind is that things change, that there is progress in almost all parts of life. Progress is a strange beast, though. Freedom of expression, advances in human rights and democracy that are taken for granted in parts of the world, take hold tenuously elsewhere and sputter to an end soon after in places where progress is considered differently. Materially, on the one hand commodities are generally cheaper than they were, while on the other hand nothing lasts as long as it used to. That applies to gadgetry, shoes and clothes, as well as relationships, news reports, hairstyles, attention spans. If things were once made to last, it is a mark of what is considered progress that they are now made purposefully to be thrown away after a short time. Darning moth holes in sweaters or holes in socks, as was once common in households, is now not imaginable. It's not worth the time and the effort, since we are chronically short of the former and are taught not to go after the latter. Perhaps it's indicative of the urge to grasp 'now,' with the commensurate feeling that the past isn't worth preserving.

Fate ('it is what it is')

It's time we look again at fate, touched upon in the first chapter. Fate is an interesting topic, since it contains so much loaded information and beliefs within a notion that is widely accepted as relevantly real. Within fate—that things were determined to happen the way they were supposed to—there's the soft version and the hard version of determinism. The soft version is that whatever happened had to happen, and that whatever will happen will have to happen: the inevitability of all events. The hard version has the added mystical element of fulfilling one's destiny: everything happens for a reason outside of us.

As good an example as one can get of fate seen as fulfilling a very personal destiny are the words of former US president George H. W. Bush, who narrowly escaped being beheaded and eaten by Japanese soldiers when he was shot down over the Pacific in the Second World War. He recalled for a CNN documentary in 2017[124] that while on the submarine that rescued him, he asked himself why he had survived. "Why had I been spared and what did God have in store for me? In my own view there's got to be some kind of destiny and I was being spared for something on Earth."

While not everyone who believes in fate necessarily also believes that heaven has a specific interest in what happens to them, or that the universe particularly cares (or can care), most believers in fate seem to at least subscribe to fate in a fairly weak, but strongly felt, form. So, for example, the thought, "If I had just left a minute earlier, I would not have been at the crossroads and so would not have been involved in the accident," or "...we would never have met," is natural.

For those who believe in fate, and even those who are skeptical about it, when something monumental happens to them, it is an overwhelming feeling that no logical explanation can shake. David Levithan's character A. in *Every Day* refers to falling in love as a sensation that could only be the result of countless years that led up to that moment, that this powerful moment is far too auspicious for it not to have happened at that time:

> ... The moment you fall in love feels like it has centuries behind it, generations—all of them rearranging themselves so that this precise, remarkable intersection could happen. In your heart, in your bones, no matter how silly you know it is, you feel that everything has been leading to this, all the secret arrows were pointing here, the universe and time itself crafted this long ago, and you are just now realizing it, you are now just arriving at the place you were always meant to be.[125]

Believing in fate is a win-win situation, though. The inevitability of fate is met with equanimity, summed up in the oft-used phrase 'it is what it is,' a modern cliché, comforting in its shrugged blandness. So 'what happens, happens,' or, about one's demise, 'when it's my time...' or 'when my time is up...' are popular and seemingly sensible down-to-earth truisms, but no more than a belief in the helplessness inherent in fate, that of 'being where I'm supposed to be.'

The idea that fate is simply what is meant to be, is comfortable and relaxing. Already mentioned is John Lennon's, "There's nowhere you can be that isn't where you're meant to be," while Lemony Snicket in the popular children's book saga *A Series of Unfortunate Events* suggests that "Fate is like a strange, unpopular restaurant filled with odd little waiters who bring you things you never asked for and don't always like."

The way that fate *always* works is that what has happened was bound to happen. With hindsight being twenty-twenty, whatever occurs will always have been fate, whichever way it turns out. It is where we are when we get to what was the future, looking back at what had been leading up to it. So when Archduke Franz Ferdinand of Austria was shot by a mentally unstable man named Gavrilo Princip in 1918, it was a cause for the start of the First World War. On the other hand, when Claus von Stauffenberg failed to kill Hitler in 1944, the war continued and millions more died. Two different events with different outcomes, but both fated.

The future and fate are imprisoned in the same box: they can be seen to happen only in retrospect. And for believers in the strong version, that of a mystical-type personal destiny, fate is justified in that its results are proof of its purpose. Psychologically, this is a necessary function. Purpose gives a sense of direction to life, that there is a reason for our existence, not just for us as individuals, but also for mankind.

The underlying conviction is that existence itself has a goal, that the universe has a goal and that we can discern the world moving onward toward that goal in the future. On the surface, this is an appealing argument. There is considerable optimism involved, since the ultimate goal of progress is perfection.

The notion of time progressing toward ultimate perfection was taken on board by Charles Darwin, who saw it as the goal of his theory of evolution. This aspect of evolution has been buried and largely forgotten by later interpreters of Darwin, who rejected its religious connotation, so that the emphasis is natural selection and 'survival of the fittest' as the ability of organisms to best survive in their environment.

In retrospect, it can seem as though history does have goals within the seeming inevitability of its progression, But this may simply be the thread—the 'plot'—that runs through any story being related. Looking for a pattern in events is a very human habit, and it is humans who give meaning to the universe (in large numbers through interpretation of the will of deities), to their sense of time and to goals that make personal existence worthwhile. Deciding whether the plot was predestined is the thin line, fuzzy in the extreme, that distinguishes believers in fate. Without the notion of fate, it would be hard sometimes to remember that, as astrophysicist and science communicator Neil deGrasse Tyson put it, "The Universe is under no obligation to make sense to you."

Believing in fate necessarily includes the assumption that it is linked to what is yet to come. Yet, it is a future gleaned from backward causality—a retroactive explanation as to future expectations. A sort of wishful thinking.

For a large number of people, believing in fate is intuitive. Their point of view is that nothing happens by chance and that all coincidences have meaning. There are no coincidences, in other words. This has repercussions on how time is perceived, since if there is meaning and purpose to every event, time itself would seem to have a purpose. Time with an attitude! With time as change being an integral part of nature, fatalists brush closely to mysticism in their outlook.

From mysticism to the paranormal, the gap is not wide. It was Carl Jung in the early to mid 19th century who delved into 'meaningful coincidences,' coining the word 'synchronicity' to describe events that occur without cause but that are 'meaningfully related,' thereby justifying a link to the paranormal. It is just a small step from that to accepting that there are things we simply don't understand but that are 'true.' Jung explained that when there are phenomena we don't understand, it puts paid to causal reasons.

But let's not forget that anything that is meaningfully related can only be seen as such retroactively, as also occurs with coincidences. Coincidences, like fate, are in a win-win situation.

Yet, despite our reasoning, we are all sometimes struck by coincidences that shake us to the core. Having the same number coming twice in a row in roulette, or bumping into your neighbor crossing the road thousands of miles from home, or someone being struck by lightning three times may be statistically possible, but nevertheless leave us quite aghast. It does sometimes seem that time works in mysterious (if not miraculous) ways, bringing events together that make us wonder if there are things in the universe that we simply don't understand. It doesn't, though. In the cold light of day, time doesn't work for us or against us, or, in fact, fit into what we consider to be reasonable behavior. Rare events and seeming miracles happen every day, according to the statistician David Hand, whose Improbability Principle explains why we should not be surprised by coincidences.[126]

What are coincidences, though? Without any apparent causal connection or explanation, mystical or otherwise, coincidences are remarkable concurrent completely random events. Randomness precludes the possibility of choice, since there is no cause and effect involved, a tough proposition to swallow within any discussion of time. For a start, how can one prove that anything occurring randomly does not have a cause? We can, surely, claim only that something *was* random. We can't observe an instance of randomness, because there is no way of recognizing an event as random compared to it being non-random. Like fate, it seems to have worked, but it seems so only due to the circular belief that it has.

Free will

If everything is meant to be, then fate and determinism go hand in hand. Fate, which is based on the idea that things are bound to happen in the way they did, and free will, which is our ability to choose our paths in life, do not, however, go hand in hand. After all, how could we make a choice about events that are meant to be, and that are meant to be every time? With fate, it would mean that the future is already determined; it's just a matter of getting there and hence no choice is involved.

That we can have no choice is a hard argument to swallow, being so obviously counterintuitive. The feeling that we have choice, that we have free will, is extremely strong. After all, we make choices all the time. We can choose how we spend our time, can we not? It certainly feels like it. Yet, can one reconcile choice with a belief in fate, which insists that everything was meant to get to where it got to at any point in time? Surely our feeling that we had a choice is shown to be unjustified when looked back from the point of view of fate.

The feeling of powerlessness at the impossibility of changing the future, of having the choice to do so, is manifested in our day-to-day language whenever we use 'if.' It is also encountered in self-help books, in the sayings of New Age and the teaching of Buddhism. "I am not what happened to me; I am what I choose to become," Carl Jung railed, reflecting views similar to the teachings of Buddhism in which he was interested, that self-development is 'mind led.' As inspiring as that sounds, it goes directly against the notion that we are the sum of our experiences ("I am not what happened to me," Jung said), and that the past influences our behavior and fashions who we are. It tells us that the future is up to us if we want it to be, and that positive thoughts and persistence can make it happen. As an early precursor of *The Secret*, mentioned in chapter 1, which preaches a certain form of extreme capitalism in which we can achieve anything we set our mind to, the insinuated message is that it is our fault if we don't succeed. "Anything can be achieved. Anybody can achieve great things in their lives if they are willing to work hard, make sacrifices, and dedicate themselves to the dream they have," said Chris Hoy, winner of the cycling gold medal at the London Olympics 2012, in an interview.[127]

Whether it is referred to as 'the power of the mind' or the more gentle, mystical 'self-realization' of Buddhism and various forms of yoga, the idea is similar: the possibility of influencing time and the free will to do so. Sri Mrinalini Mata talks about "the infinite potential of the mind," the cultivation of willpower and how the practice of Kriya Yoga can uplift and transform people's lives.[128] The dark side of the so-called power of the mind that can affect the future is exemplified by Rhonda Byrne, author of *The Secret*, who said that the victims of the tsusami in 2006 were sending out tsusami-like signals into the universe. The notion that the energy generated by the mind has the potential to affect other energies and the material world is demonstrated spectacularly in the 2014 movie *Lucy*, in which the main protagonist's mind is unlocked to its full potential through a super drug, so that she can understand and therefore control time, in order to affect the workings of the universe.

A consequence of being able to control time would mean knowing the future. The principle, emanating from Newton in 1687 and expounded upon by Pierre-Simon Laplace in 1814, is deceptively simple: if we knew everything about the present, then we could predict precisely the future. That notion is problematic, though, if we consider the fact that we are never in the present and so cannot know *anything* about it, never mind *everything* about it.[129]

Mapping the past in order to predict the future might seem more promising. Dr Kira Radinsky, an Israeli computer scientist, developed a technique for predicting events with an 80 percent average rate of accuracy by scanning literature written during the past 500 years, including everything published in *The New York Times* since 1880. Radinsky discovered that integrating information on famines and heavy storms was a way of predicting the outbreak of cholera. And, yes, 80 percent accuracy is very high. But it was accurate about what *had* happened after being predicted, not about what *will* happen. Cause and effect about past events need to be tested about future events, i.e. events that have not yet happened—an inherent impossibility.

An argument for free will is not only that we are able to affect the future, but also that we could have chosen something else rather than what we did choose. The implication is that there are numerous possibilities for us to choose from, or as British philosopher Julian Baggini puts it, "In the shadow of every life led are all our potential lives not led."[130] If, for example, we decided to cross the road, we could instead have decided not to do so. But this doesn't show anything about exercising free will since, like fate, we are looking only after the event at what action we chose. That does not stop us fantasizing about it, though. In a fervent wish for retroactive free will through time travel to the past, Italo Calvino in his novel *If on a Winter's Night a Traveler* says, "This is what I mean when I say I would like to swim against the stream of time: I would like to erase the consequences of certain events and restore an initial condition."

Yet, unlike the plot of a novel or a movie or a TV series, our actions cannot be unwound in order to change them if we wish. In the framework of choice within the course of time, looking back is no different from looking forward: there were no more possibilities to choose in the past than there are in the future. Fate being fixed does not allow either, while free will would allow both. When historian Yuval Noah Harari tells us that when we study history, we do so "to understand that our present situation is neither natural nor inevitable, and that we consequently have many more possibilities before us than we imagine,"[131] he is subscribing to retroactive free will. How, from a position *after events have happened,* could our location in time not be natural or inevitable? Surely, there can be nothing more natural than where we are. The line between fate and free will can, indeed, be fuzzy.

Strange it is that those who believe in fate are so calm about it, so content in the knowledge that *che sará, sará* (what will be, will be), while also holding onto the belief that they have free will. Instead, ways of having the cake (fate) and eating it (free will) are inventive. So while New Age is largely prone toward the comfort of fate, it is also big on the fact that chance and luck can, and does, change one's life. The best plans can go wrong, they say, when a chance occurrence comes along. In such a possible scenario, free will is not that free, surely. "It isn't enough to just plan your future," the website The Tree of Awakening tells us. "You've got to have chance on your side. In fact you must have chance on your side. Without it, whatever you do will go wrong." How does it allow free will, as well? If every event, however small, could cause effects everywhere—the butterfly effect—surely that shows the effects cannot be anything but a cause for something else to happen.

Also Buddhism has enigmatic views about choice. We are told that "We can reflect on past events and learn from them, but we cannot influence them. The past has a fixity that contrasts sharply with the more malleable future, where we make choices and influence events."[132] As we have seen, though, the past is far from being fixed as far as humans are concerned, while the future is not at all flexible when seen in retrospect as having been fated to happen. Furthermore, with Buddhism's deep belief in a strict interdependence of all things, it makes us wonder how it can also advocate free will and choice. With all of us being dependent upon everyone and everything else, surely mutual influence—cause and effect—is paramount. How and where is choice in this mixture?

Not many people would argue that free will is not worth having. Most of us would like the idea of personal freedom. But an argument based solely on what feels good is not much of an argument. The proof is in the pudding, and you always end up with the pudding whatever you do:

> Alice came to a fork in the road. "Which road do I take?" she asked.
> "Where do you want to go?" responded the Cheshire Cat.
> "I don't know," Alice answered.
> "Then," said the Cat, "it doesn't matter."
>
> (Lewis Carroll, *Alice in Wonderland*)

There is no such thing as 'a road not taken' when it refers to the past. Both free will and fate are always about the road that *was* taken. Faced with a decision about the future, all roads are not yet taken.

Whether we call it fate, predestination, destiny, determinism, foreknowledge, prescience, preordination or predetermination, the results are unknown until they happen. We can add astrology, horoscopes and reading tea leaves to the mixture and we will get the same result. This may be sad, but it is an unavoidable fact. We cannot get to the future, whichever term we use and however much we believe in a particular ability or way to do so.

In all of this, it's a matter of perception and manipulation of language—a way of sugaring the pill according to one's personal beliefs. To some, it sounds somehow much gentler and more reasonable to believe in *destiny* rather than *determinism* or in *fatalism* rather than *predestination*. All advocate cause and effect, that what happens is the result of past actions. It's very subjective, of course, and it depends upon your audience. Churchgoers are more likely to accept with equanimity that life is *preordained* rather than it is *predetermined,* even though it has a similar result. But using the word *preordained* in a discussion of philosophers would be unacceptable. Whether we think we are powerless to change events, or that there is a cause to every effect, the difference is a matter of nuance and belief. After all, even the surest fatalist will watch out for what is coming before stepping into the road, as will the predeterminist and the believer in free will. If there is an accident despite their caution, each will claim that it was meant to be.

The issue of fate/predestination and free will ties itself into knots when religions are brought in, where there is a supernatural being, having an existence that is not within time, and enabling that being, unlike humans, to genuinely predict the future. It certainly is a nice way out of the conundrum of prediction: God knows what is coming, even if we don't.

The principle of free will in religion should be simple: if God knows what we are going to do, then our fate is determined and our actions are predetermined. In this case, we cannot have free will. It makes no difference if we feel we have choice at any time. Once our future is known—really and definitely known, as can apply only to God—nothing we can do will change it. So as to make it absolutely clear, we are referring to God as an all-knowing, all-powerful entity. Unlike humans, God knows what we are going to do. If he doesn't know, he is not all-knowing.

Religions, though, claim that God has given humans free will, which is a necessary belief within a system that contains culpability for sin. After all, there would be no original sin, no punishment by God of Adam and Eve, if they didn't have free will to disobey God's explicit command not to eat from the tree of life. That is hugely problematic, since it contradicts the fact of

God's knowing what we will do. When American philosopher and psychologist William James famously remarked that the future is so unreal that even God cannot anticipate it, he went overboard with that provocative statement. For a start, to say that there is something that God cannot do is to deny the powers of God. Secondly, it demonstrates a not-uncommon confusion between the future as something that we are moving toward and details of what that future will contain. The fact that we obviously don't know what will be, does not contradict our knowing that there is a future.

This is a sensitive point. In fact, any argument against free will is sensitive and often provokes an emotional response, with or without God in the mixture. Generally, we object to being told that we can't decide our actions. The argument that you still have free will even if someone (not necessarily God) knows what you will do, does not hold water. In other words, if I claim to know what you are going to do, it doesn't mean that you have to do it. Of course you don't. You can also change your mind at the last minute.

Take the book you are reading at this moment. Obviously, I know what is printed on the pages that follow, thus demonstrating that I know the future as far as you are concerned. But do I? If you decide to stop reading, I would not have known that beforehand, since you would not have reached the part I thought was your future. It turns out, then, that I know the future only of the things I think you will do, not of what I don't know you will do. In other words, I cannot know what you will definitely do. I, unlike God, am within time and am not infallible; I cannot be 100 percent right about what will happen. We cannot get away from the fact that the future is where we never get to.

That the future always pans out in retrospect is inherently so also for trends. As a means of pointing to the future, trends are little more than speculation based on the past, and wrapped up in the aura of science and statistics. They are a way of forming patterns or, as the evolutionary biologist Dominic Johnson put it, "Our brains are wired such that we cannot help but search for meaning in the randomness of life." Trends are always correct. They can be faulted only when they have ceased predicting but are instead the past.

The 20th century was going to be the electric century, with the streets of the world filled with electric cars. So it seemed when the first electric car was invented in 1837, which led to a fleet of electric taxis in London at the beginning of the 19th century. A clear trend. But, as we know, it evaporated. Instead, we have the bane of the noisier, polluting gasoline-driven vehicles.[133] And now that we are re-trending electric vehicles and also driverless cars, we feel again the certainty that they are the definite future.

So many trends have not come to pass or, if they did, the results were sporadic and unreliable. The price of oil in the mid 20th century looked as though it would continue to rise forever, only to have taken a nosedive in the first decade of the 21st century. Anyone who speculated on gold or tin, or indeed a number of natural materials, was battered when following what looked like trends. Or China, which looked certain to obtain world economic domination. Trends that have stopped, prove themselves not to have been trends, at all.

Yet, we thrive on trends and continue to believe the continuation that they promise. Much of the world's economy rides on trends, in many cases the engine that drives it all forward. Where would the fashion industry be without its trendy people to show us what we should wear? 'Fashion' is per se a transient notion; its very existence is determined by periodic change and as frequently as possible. 'Moving with the times' is the modern version of 'keeping up with the Joneses,' as a way of looking back in order to be pushed toward the future, where things are obviously better than they are today. Bankers look back and show that the long-term trends have always been for rising markets. The advice is based on past performance, which has 'in the long run' shown an increase. But that does not necessarily mean that it will happen in the future.

'Moving with the times' is rather a chicken-and-egg situation. It makes no difference which came, or comes, before the other. Whether we affect time (setting a trend) or time affects us (following a trend), we again have our metaphor of circular time that moves forward toward the future.

Knowledge of future events would invalidate free will. It's why we don't—and can't—know the future; if we did, we would not even have the illusion of our having free will. Free will depends on our belief in a future that we can affect, even if we don't know what the future will bring.

It goes without saying that that there can be no notion of a future that has no past. It seems strange, then, that there is a perplexing question mulled over widely by philosophers and more recently by scientists, which seems to imagine such a possibility (actually, the possibility of such a possibility). The question is: *Why is there something rather than nothing?*

With apologies to anyone I might offend, the question as to why there is something rather than nothing, is, I suggest, not the profound question that Martin Heidegger famously characterized as the most fundamental issue of philosophy. Or what that philosopher Bede Rundle calls "philosophy's central, and most perplexing, question."

There is nothing astonishing for there to be a world, for nature to exist and for us to be here. When author Jim Holt asks: "Why does the world exist?" in the book of the same name, the answer surely is that it couldn't be any other way. Even if scientists could give explanations about how the universe was formed from Nothing, the 'perplexing question' would still not need to be asked. If the universe didn't exist—if there had always been Nothing—there would be no one to answer the question, let alone ask it or, in fact, to read Holt's reasons for asking it. Asking 'why' in this case is, in fact, one of those speculative anything-can-happen 'if' questions. Having already arrived at what was once the future, the questioner asks what would have happened if it had not happened! What could reality be other than what there is? What sort of 'reality' could there be in a world with nothing in it (whatever that could possibly mean)? Do scientists, perhaps, theorize such a situation to mean a world in which there is nothing apart from humans? Would there be no time in such a world? Would there be change if there wasn't anything that changed?

The bottom line is that if the universe weren't the way it is, we wouldn't be around to observe it. The question 'Why is there something rather than nothing?' is, in other words, itself empty. There would be nobody to answer the question.

Without appeal to a higher being ('God wanted there to be something'), the question as to why there is something rather than nothing cannot have an answer, even a speculative one. It is not a matter of why or how the universe was created; the question is: What would be if it had never come into existence? With the absence of a past, and no lead-up to what there is, there wouldn't be anyone to ask the question and there would be no into-the-past moving *now*. It would be asking how we got to where we are without getting here.

IV

Dissonance, Fragmentation, Approximation

Change

In our foray into time and its effects, a striking phenomenon became apparent as we moved along. We kept on coming across stops and starts, paradoxes and puzzles strewn along our path; in a nutshell: tension underlying harmony; splits rather than unity; good enough rather than spot-on. It shouldn't be surprising. Time disturbs what would otherwise be nothing. It would not be hyperbole to state that time is at the basis of nature itself. Time, after all, is change: everything always and constantly changes, so that the very movement of time is interference-causing tension.

There is nothing static in nature; all is in movement, everything is always happening. It doesn't matter how steady and 'uninteresting' a town, village or country seems to be, newspapers are full of news. Never is there a headline 'Nothing happened.'

Time is what makes sense of the observable, because its inherent contradictions mirror those within humans themselves. Already some 1500 years ago, Augustine noted the inner tensions that are an intricate part of humans, when he wrote that "Time is nothing other than tension, and I would be very surprised if it is not tension in consciousness itself."

Dissonance is in the very nature of time, manifested in the constantly backward-moving *now*. It is this *now* that is the key to the shifting, seeming erratically moving time, allowing us to make sense of our world.

It is important to understand what this moving *now* is. Or rather, and particularly, what it is not. *Now* is not a thing. As we have seen, *now* is not a period in time. It cannot be measured. And it cannot be predicted. Because it does not have a clear and defined trajectory into the past, there is no way of knowing exactly when it will become memory or even if it will be a genuine memory at all. It is because *now* is constantly moving that memory is not certain but is a range of possibilities within the past, projecting a range of possibilities for the future. How we interpret *now* in our day-to-day world is up to us, so that 'the present' could be an hour, a day, a year, or whatever we take as a particular context. But what it is that we are interpreting is the moving *now*, which we could imagine metaphorically as a fundamental and random continual bubbling of memory formation, and would explain the vagaries of human memory as part the individual's own psychological make-up.

The beauty of the moving *now* is that it gives an explanation of how things work in time—in fact, how time works. It is a logical description of the impossibility of *now* and *the present* as a part of time, and that its untouchability, immeasurability and indefiniteness is the reason for the vicissitudes of the past and future, and the continuously changing self.

While there are others, including, of course, New Agers and Eastern mystics who consider *now* as an especially significant presence, and regard it—even when referring to a *moving present*—as a period in time.[134] I do not! We have demonstrated that 'now' (the present) is not a period in time. *Now* is 'real,' in the sense that it has effects: it is observed through what it does. So, for example, change, the result of the backward-moving *now*, can be seen only after it has taken place.

Many of the phenomena that we have come across are explained through that probabilistic *now*. It would explain why, for example, the future is a range of possible or probable events based on an uncertain past and so cannot be one of prediction; the probablistic moving *now* is the reason we can never alight on the future. And it explains why change is always discerned after it has happened.

It might be a good idea at this juncture to have a breather

and to look again at what we could possibly mean by 'real,' in particular when it comes to 'now' and 'the present.' It is especially interesting concerning the moving *now*, **the results of which** are 'real' in the sense that they are part of our day-to-day world. The moving *now* itself, though, is only felt but never seen; nor is its exact position known.

One way of dealing with 'real' could be to refer to what is **consciously real** (that which we feel and perceive as real) and what is **theoretically real** (an explanation as to what may be behind the consciously real). But we won't do that, since it implies that what we perceive in the 'real world' is less 'real' to us than what might be behind it. We see things the way our brain interprets them, so that to us, they are 'real.' As to what anything 'theoretically really is,' all we can do is conjecture, theorize and imagine.

When it comes to the moving *now*, though, we are in a bind. We can't track it or see it in any form. How do we refer to *now* as 'real' in any way, when the evidence we have of its appearance is only its effect?

We have to leave a question mark on this discussion. We can only theorize about what may be 'really happening' behind the scenes of what we perceive. Any answers we come up with will always be hidden from us, limited as we are by what our brain tells us is happening. Very apt is the exchange in the TV science fiction series *Westworld*, in which the protagonist asks the hostess, "Are you real?", to which she replies, "If you can't tell, does it matter?"

Dissonance

"The universe is not required to be in perfect harmony with human ambition."
Carl Sagan.

It's an interesting statement, and makes one wonder why so

much time and energy is spent on aiming for the unattainable mirage of human harmony with nature. There's more to it, though. Behind it lies the question as to whether the universe—nature—is itself in perfect harmony. It certainly looks as though it is, with everything seemingly working in a balanced clockwork fashion and all aligned just right. It's not quite as good as it looks, though. The very nature of time as change indicates an underlying state of constant self-righting tension.

Poised on a critical edge between what is observed and what is about to happen, systems—all life, in fact—are constantly on the verge of change, metaphorically like a pile of sand that we keep adding to. At some point the pile of sand will partially collapse and reform to form another pile. Basically, the tension is between *being* and *being something else*, between *order* and *disorder*, the latter always being held at bay by the urge to stay the same. It's a delicate line, the dissonance that hold things together close enough for them to function.

With dissonance so ubiquitous, and perfect harmony a seeming impossibility, it is dissonance, fragmentation, and approximation that encompass the description of how nature works.

Until the early 20th century, Newton's view of a universe that works in an orderly fashion was rooted within culture in all spheres. Order was behind everything and meant predictability. In the meantime, Newton's ordered universe has been augmented by quantum mechanics, which applies to subatomic particles, the building blocks of nature, and shows that these work in unpredictable ways. Interestingly, the basic nature of all things has been shown to be anything but 'smooth,' where the very nature of space itself is granular. Reflecting our observation in the previous chapter that any line, however straight it appears to be, has a fuzzy edge the closer you examine it, physicist Carlo Rovelli refers to quantum mechanics, when he says it "reveals to us that, the more we look at the detail of the world, the less constant it is."[135] This is not surprising when we consider time, constituting the very heart of nature, as containing the constantly backward-moving *now* as its basic manifestation.

Some 70 years ago, the Austrian Nobel prizewinning physicist and quantum pioneer Erwin Schrödinger suggested in his book *What is Life?* the paradoxical fact that the laws of classical physics, such as those of Newtonian mechanics and thermodynamics, are ultimately based on disorder. Stephen Hawking refers to "the luck of imperfection" when the universe was formed after the Big Bang through gravity. He maintains that there is no such thing as perfection. Had there been perfection, he says, the stars would have aligned themselves perfectly, equidistant from each other, and nothing would have happened. Instead, with slight imperfections, the stars came together in clusters.[136]

Long, long before modern science, Heraclitus, the Greek philosopher who was active around 500 BC, propounded a distinctive theory that things are constantly changing (universal flux), that opposites coincide (unity of opposites). Quite brilliant that was, in fact, considering how far down the road science still was, and what a leap in time and knowledge there was compared to what we know about the subatomic elements described through quantum mechanics.

How could, in fact, nature be harmonious when humans, themselves a mess of dissonance, are part of nature? Infused with doubt, we have an incurable urge to seek remedies. Lest it be thought otherwise, doubt is not a modern phenomenon. "We are the species that discovered doubt," says American philosopher and cognitive scientist Daniel C. Dennett. "Is there enough food laid by for winter? Have I miscalculated? Is my mate cheating on me? Should we have moved south? Is it safe to enter this cave?... It is that deep-seated doubt that causes an almost schizophrenic reaction: doubt vs. truth, reality, as a primeval dissonance."[137]

Humans are full of contradictions, contradictory thoughts and feelings. We've come across many of those in our foray though time. The fact that everything is relative to everything else, is itself a path though differing sides. Whatever we think about is set off against, and compared with, something else, just as every physical object is defined by its surrounding space.

People believe and act in contradictory ways without getting too hung up about it. There are scientists who spend their days experimenting toward evidential proofs, yet come home and believe in a supernatural being who created the universe. There are medical practitioners who smoke, soldiers who believe in the sanctity of life and kill when they are ordered to, people who love animals yet eat meat, and brutal dictators who can still be loving fathers and care for their dog. A reason for these internal contradictions, according to anthropologist David Berliner, could be compartmentalization, where contradictory ideas live side by side and are released to suit a current situation.[138]

And that's the point: current situations are always temporary. They are of necessity temporary, since people cannot act out conflicting reactions simultaneously. Nevertheless, holding onto conflicting ideas, beliefs, values or emotional reactions within the same person—with the term 'cognitive dissonance' coined by Leon Festinger in 1956 to describe the condition—would seem to be part of human nature. The fact that people look for ways to reduce dissonance in order to find a balance is normal; yet its aim is not only impossible to achieve, but can itself sharpen the dissonance.

It happens often on the fringes of religious believers. In a similar way to extreme orthodox Jews in Israel and elsewhere who eschew modern clothes, Amish groups, including the Mennonites and the Brethren, living in Lancaster and Dutch County, Pennsylvania continue an even more cloistered lifestyle, with their plain black-and-white clothing and horse-drawn carriages. Among these stricter groups, electricity of all kinds—

phones, televisions, radios and cars—are forbidden in a supreme effort at keeping out the modern world. Yet, even here time has not stopped when it comes to keeping up with earning a living, so that, apart from farming, it is not uncommon to see Amish-owned shops and even well-appointed hotels with all mod cons.

Dissonance is a state of the world as it functions, where the backward-moving *now* keeps contradictory forces in a state of a seeming working balance, an approximate balance that we could term 'dissonant balance.' The obsession with trying to achieve perfect harmony—and this *is* a modern phenomenon—is misplaced.

Yet, the notion that there is symmetry in nature is deeply satisfying. It feels as though that is how things should be. And, in fact, that is how nature appears to work. Most often, though, symmetry itself springs from asymmetry, from a basic dissonance that ultimately holds things together—the universe in a state of dissonant balance.

But not all is symmetry. Our bodies are not symmetrical; one side is different from the other. Each eye may be a slightly different color. One shoulder may droop, as may one of our eyelids. Try walking in a straight line with your eyes closed and you will notice that symmetry is not a natural state of affairs. No tree or bush is symmetric; nor is virtually anything in nature. Approximate symmetry is where we can get to. Folding a piece of paper down the middle will produce two seemingly equal sides, with close examination showing that they are not exactly symmetrical.

It would be surprising if anything were exactly symmetrical, if only for the fact that change is ubiquitous and continuous. It is for that reason also that beauty is not necessarily a result of symmetry, and that women considered beautiful do not have symmetrical faces. More surprising is that Photoshopping a face to make it symmetrical results in weird and unfortunate results that are far from flattering. This is what Alex John Beck discovered from his photography project *Both Sides Of* in 2014, in which he matched the left side and right side of faces in a series of photos. The Ancient Greeks tried mix and match long before. They considered naked, athletic bodies to encapsulate perfection, but knowing this was unlikely to exist in the real world, their artists merged what they considered the ideal features from different bodies—often male and female—into the one form, to create what they considered the perfect specimen. We don't know what their models thought of the results, but some of the subjects of Beck's Photoshopped photography thought the results were so strange, so unattractive, in fact, that the subjects did not allow them to be part of the exhibition.

It isn't surprising that beauty and symmetry do not necessarily coincide. Beauty is an ideal, a state of perfection, that as far as Plato was concerned is a theoretical concept only. "If you look for perfection, you'll never be satisfied," said Lvov's wife in Leo Tolstoy's *Anna Karenina*. Not containing flaws, perfection is not found in nature. Attractiveness, though, is another matter. Attractiveness is individual and different in every case. Unpredictable because of its attendant flaws, it covers an almost infinite range of possibilities.

Music, like everything else, has to be flawed in order to exist as itself; as soon as it becomes perfect (through electronically forming the flawless sound), it disappears into the perfection of its own special effects.[139]

Whereas in Ancient Greek and Egyptian societies, artists created impossibly high ideals for the beautiful body, later cultures changed over time and artists gave up on the idea of achieving perfection, understanding that a degree of dissonance would give the desired results. As Leonard Cohen expressed it: "There's a crack in everything. That's how the light gets in."[140] Builders in antiquity purposely and secretly introduced minute variation from absolute symmetry in their columnar structures. In Thomas Mann's *The Magic Mountain*, the character Hans

Castorp describes drops of ice, "each of [which] was absolutely symmetrical, icily regular in form. They were too regular, as substance adapted to life never was to this degree—the living principle shuddered at this perfect precision, found it deathly, the very marrow of death."

The cliché 'beauty is in the eyes of the beholder' is about attractiveness as well as beauty, neither of which can be objective. There is no objective beauty 'out there' or even 'in there'; the symmetrical artwork representing beauty that could be produced is a patchwork of perfect physical attributes that does not exist and cannot exist in one person. "Have no fear of perfection," said Spanish surrealist artist Salvador Dali. "You'll never reach it." Somewhat akin to 'averageness,' the concept of beauty as an ideal fits no one. As for attractiveness, this is by definition a subjective point of view.

The media's emphasis on physical perfection is shorthand for supposed worth: that outer beauty is a mirror of inner beauty. As a fleeting attribute, it captures the popular perception of 'now' as the in-your-face be-all and end-all.

For science, the beauty of a theory or a mathematical formula means elegance and simplicity, and is important in itself. As controversial as that is, some of the greatest minds subscribed to the notion. When Einstein said that, "the only physical theories that we are willing to accept are the beautiful ones," he seemed to be asserting that beauty trumps the inelegant results of experimentation. Also seemingly arguing for 'beauty is truth' is Hermann Weyl, a colleague of Einstein's, who said that "My work always tries to unite the true with the beautiful; but when I had to choose one or the other, I usually chose the beautiful." Even more blatant was the Nobel prizewinning physicist Paul Dirac, saying in 1963 that "it is more important to have beauty in one's equations than to have them fit experiment."[141]

There is, though, an important difference between the use of 'beauty' as applied to celebrities and that used by scientists.

For the former, beauty is accepted as fleeting, and reflects the prevalent idea of seizing the moment and living 'now.' For science, a theory is beautiful when it gives an explanation of phenomena in a simple way; in other words, containing an aura of certainty for its efficacy in the future, it is meant to be more than fleeting.

Science itself is not about certainty, though, and nor should it be looking for it, even if scientists sometimes give the impression that they are. It is about the patient weighing of experimental results according to the current level of knowledge, which is reliable for the moment until something more reliable comes along. Knowledge, as we saw earlier, is necessarily transient; the inherent change that is time sees to that. Science moves forward by refuting theories, looking for dissonance and embracing it. As Saul Perlmutter, Nobel prizewinning astrophysicist, says: "Science isn't a matter of trying to prove something—it is a matter of trying to figure out how you are wrong and trying to find your mistakes."[142]

The temporary state of knowledge and the willingness to live with uncertainty is the epitome of dissonance: the lack of certainty that grounds it and gives the incentive for further research. Until we know everything—and we never will—what remains to be known is by its very nature dissonant with what is known today.

The very search for knowledge is the unwilling perpetuation of dissonance. The certainty of what is known today relies on the probability that it will be changed in the future. While science is never objectively true, it is in the very nature of humans to search for knowledge, to shake the existing state of affairs and to continue shaking it.

Even believers in a perfect God are faced with a universe that is far from being perfect or balanced. It all began with chaos, according to the Bible, which states that in the beginning there was '*tou vavou*,' Hebrew for 'disorder' and 'chaos.' Things didn't improve after that, with Eve not at all content in the initial perfection of the Garden of Eden. Thereafter, it was downhill all the way, with evil and general mayhem on a roller-coaster ride to the present. Perfection would be objective and absolute which, by definition would be timeless. Even God didn't succeed in preserving that original impossibility, it seems.

It would be difficult to make a case for absolute harmony in any aspect of the world. The Cold War standoff between the USA and the USSR was a frightening example of the impossibility of more than the semblance of balance that would be more than temporary. The basic idea, 'balance of power,' never happened. Both sides made sure that they had what the other side had, with each compiling an arsenal to match the other, with the world becoming an evermore dangerous place. The search for stability was inherently destabilizing. Harmony is itself unstable, resulting in dissonant balance.

It should be surprising (but somehow isn't) that in an age where the notion 'global village' is already a cliché, there is a huge rise in wanton and unbridled violence. Mass terrorism is rampant from Pakistan to Nigeria, Syria, Somalia and Sudan, with terrorist attacks evermore frequent in quiet European cities. In this global village, where the world has contracted through technology and instant movement of information, harmony is even further from consummation. With an ever-widening gap between the haves and the have-nots—resentment being engendered by the richest owning an ever-larger percentage of wealth, while the poorest grow and possess an ever-smaller percentage of resources—the very thought of harmony is risible. Author John Gray seemed to have a point when he said that "Man, much more than baboon or wolf, is an animal formed for conflict; his life seems meaningless to him without it."[143]

Everything that we have alighted upon in our path through time has been part of a tale of opposites, paradoxes and tension that sort themselves out within situations that work. William Blake in his *Marriage of Heaven and Hell* wrote that "Without contraries is no progression, attraction and repulsion, reason and energy, love and hate, are necessary to human existence."

Yet, it is harmony that is nowadays so popularly sought as the ultimate panacea. Here are the ringing words spoken by Margaret Thatcher when she became British prime minister in 1979: "Where there is discord, may we bring harmony. Where there is error, may we bring truth. Where there is doubt, may we bring faith. And where there is despair, may we bring hope."[144]

Dissonance is frowned upon when happiness is the ultimate aim. After all, the unalienable right to the pursuit of happiness is stated in the United States Declaration of Independence. Theologian and mystic Thomas Merton strikes a chord when he states that "Happiness is not a matter of intensity but of balance, order, rhythm and harmony." The notion of a simpler, balanced existence does seem reasonable. 'All you need is love' was the message of the 1960s, when 'flower power' was the way to peace and happiness, and New Age became a simple and hip way of achieving a perfect utopian balance. As fractured and diverse as New Age can be, an underlying theme is the presence of order behind the disorder that humans come across. That, though, is contrary to the modern theories of science, in which the opposite is claimed: that there is basically disorder, with a continual movement from order to disorder beneath the order that we see.

The urge to understand time, and thus to be able to control it, is endemic and the source of still more dissonance. The Foundation for Inner Peace, set up in 1975, which published their tome *A Course in Miracles*, has sold in the region of 2 million copies in 22 languages. The search for tranquility is great indeed, while the pursuit of happiness is even more so. The notion that one has to be happy may be the cause of much unhappiness, with writer and psychologist Adam Phillips maintaining that the pressure to be happy is driving people mad.

Trying to be happy has everything to do with time and making it work the way we want it to. The popular solution and advice is to slow down, to take our time. So says, for example, Haemin Sunim, the 'mega monk' author of a multimillion-copy bestseller on the subject, *The Things You Can See Only When You Slow Down*. It does seem a reasonable way of dealing with the super-fast surroundings and climate of rolling news that we can't get away from, and where there is no time for sitting and thinking.

It certainly sounds like good advice. But what is it about 'slowing down' that is particularly good? Perhaps more to the point is whether it is possible or even practicable to slow down. And what is the pervasive 'mindfulness'—a synonym for slowing down in order to grasp 'the now'—if not a system of slowing the pace of our own personal time? And that, of course, brings us back to the first chapter of this book, when we looked at how time seems to speed up when we are having fun and slows down when we are bored, for example. The point is that common (accepted, local) time does not change its rate, no matter what we do. It is our inner (subjective) time that seems to change its pace. But even subjective time will not go faster when you are thinking about it even while having fun. Mindfulness does nothing to time—neither to your subjective time nor objective time. If you succeed in slowing down when you are in a state of mindfulness, all it means is that you are doing fewer things in the time you would otherwise do more.

Whether mindfulness enhances wellbeing or is beneficial in any way is an issue on its own. And not everyone wants to achieve stress-free certainty. Louis C.K. said that he didn't take the job he was offered as a talk-show host because if he did so, he would see his whole life laid out ahead of him. He didn't want that. He needed the uncertainty of stand-up comedy over the steady path where everything is known.[145]

Progress often comes from disagreement, not from agreement; from failure, not from success. Warren Buffet, who controls Coca-Cola, at its annual meeting in April 2013, stated that he liked to study failure. "We want to see what has caused business to go bad, and the biggest thing that kills them is complacency. You want a restlessness. A feeling that somebody's always after you, but you're going to stay ahead." Restlessness and a constant look toward the future.

Progress emerges as a result of—in fact, a clash between—what we don't know and what we do know. Progress would stop if it were only a matter of what we know. It can work only when it takes into account that which we still don't know. There was stagnation in the Middle Ages because people had to agree with the Church. From the Renaissance onward, there was a tremendous surge in all fields, not because of agreement, but because of argument, challenging earlier ideas and the willingness to go out on a limb knowing that one's ideas would be challenged. Here is how Orson Welles' character Harry Lime put it in the movie *The Third Man* (1949): "In Italy, for thirty years under the Borgias, they had warfare, terror, murder and bloodshed, but they produced Michelangelo, Leonardo da Vinci and the Renaissance. In Switzerland, they had brotherly love, they had five hundred years of democracy and peace—and what did that produce? The cuckoo clock."

The continual need for scientists and philosophers to defend their theories is the healthy perpetuation of dissonance. If we can't ask questions and be skeptical about common assumptions, we leave ourselves open to be manipulated by those who supposedly know.

Whether balance/harmony is an illusion or a delusion is a moot point. That it is something to be desired is almost universally taken for granted. The popular notion is that coherence, balance and harmony is the ideal, and that ideal relationships are fostered within complete harmony. If this is what people expect, it's no wonder that so many marriages fall apart. Yet the search for balance, in which the permanent 'now' is accepted as a part, is natural and relentless. And so it is with the feeling—pumped up in newspapers, magazine articles and television talk shows—that work-life balance is what should be aimed for, that simply adds to the feelings of inadequacy and guilt of both men and women.

The problem is that true balance cannot be found, since it is an impossibility. Life is movement and change, so that the effort to find what we feel life should be is thwarted, since the search itself is part of that change. It was Darwin who said it's not the strongest of the species that survives, nor the most intelligent, but the one most responsive to change. Balance is movement within the approximation of an oscillating string. Reminiscent of a string of a violin and the sound of a musical note, balance can only be movement: a sound that has no vibrato is lifelessly flat.

'All in one rhythm' was the official slogan of the 2014 world soccer cup. So much in one rhythm that the slogan, being a registered trademark, actually meant that you could be only in the rhythm decided upon by the organizers.

Approximation

With all that we have discovered in our discussions, we can but ask an important question: How do we thread our acts together into a narrative that makes sense to us as individuals and how do we do it within a common narrative that allows us to interact with others? In short, how do we function through the dissonance that is endemic in humans? How do we get along in the world, when with different types of sensory information (sound, sight, touch and more) coming at us, our brains have to process them at different speeds by different neural structures? To that must be added the significance of information streaming in at varying speeds from different locations, with each person living in a (slightly) different time from everyone else. If the brain didn't sort all of that out, it would all be a cacophony of incoherent sights, sounds and a host of other senses.

Yet we deal with a world in which everything is approximate. We know that events cannot occur at the same time and we also know why: they take place in different locations and hence must be at different times in relation to where the observer is. When an orchestra begins and ends exactly together, it sounds that way to us. But, in fact, there are measurable differences—albeit extremely small—in the timings of all the instruments. Because the differences are too small to be discerned by the human ear (unless it's an extremely bad orchestra) in a concert hall, what we hear is a perfectly timed performance.

On a person-to-person level, Graham Greene posed a question and answered it in his way: "Time has its revenges, but revenge seems so often sour. Wouldn't we all do better not trying to understand, accepting the fact that no human being will ever understand another, not a wife with a husband, nor a parent a child?"

Yet, we *do* understand each other. Approximately. Well enough for day-to-day communication. Never 100 percent, but generally enough. We get along with others not by looking for balance and searching out unrealistic harmony. Rather, it is understanding that there is dissonance and that it needs to be accepted, even embraced. Accepting what is, is more productive than striving for what can never be found. Once we accept that we are not in perfect balance with others but that we are relative to others, we can relax within that instability.

Colors. Let's have a look. Apart from people who are colorblind, the majority of us can point to red and agree that it is red. But we as individuals don't know how anyone else sees that red exactly. We can't know. Whether it is seen by someone as lighter or darker (according to how their brain interprets the light wavelength hitting their eyes), we will agree on 'red' to describe what we each see because it is close enough for it not to matter. Like language—close enough for us to understand others.

But it's not just colors. It's everything. The world works—and works well—because of approximation. As the 19th-century English philosopher John Stuart Mill put it, "There is no such thing as absolute certainty, but there is assurance sufficient for the purposes of human life." There is no real or proper, objective, color red, just as there is no real, objective memory. Or a real, objective anything. What we have, then, is approximation, which is part of movement and constant change.

This is not to say that dictionaries aren't useful. They are, though, limited to what we know already. Let's try this with, say, 'elephant':

> A thickset, usually extremely large, nearly hairless, herbivorous mammal (family Elephantidae, the elephant family) that has a snout elongated into a muscular trunk and two incisors in the upper jaw developed especially in the male into long ivory tusks.
> (Merriam-Webster)

> Either of two large, five-toed pachyderms of the family Elephantidae, characterized by a long, prehensile trunk formed of the nose and upper lip, including Loxodonta africana (African elephant) with enormous flapping ears, two fingerlike projections at the end of the trunk, and ivory tusks, and Elephas maximus (Indian elephant) with smaller ears, one projection at the end of the trunk, and ivory tusks almost exclusively in males.
> (dictionary.com)

Despite the many words, they don't do the job since, for a start, we need to look up many of those words (the important ones!) in the definitions. So let's cut it down to 'an elephant is a pachyderm with tusks.' But that doesn't work either, since we would have to know (and so look up) what a 'pachyderm' is. If we cut it down further to 'a big animal with tusks,' we are getting closer to a meaning that we can understand. We would, of course, need to know what tusks are. And 'big animal' is so vague that we still haven't got a meaning that is exact (considering that a number of other big animals have tusks, including the walrus, the rhinoceros whose tusks are called horns and the hippopotamus whose incisor teeth are considered tusks). In other words, meanings are necessarily digressive and circular, always requiring the understanding of other meanings that are being used in the definitions.

It seems, then, that approximation is the name of the game and that the more exact you get, the less is understood. This turns out to be great for lawyers, who make their living from attempting the opposite. When they draw up contracts, every word is weighed and agreed upon, so that when a contract is signed, all sides are sure they know what it means. All is well and good; but if problems arise later—be they personal or commercial—it's back to the contract, which suddenly is put under a microscope because both sides no longer agree on what the words and phrases mean. Subsequently in court, it is the judge who decides what the meaning is—possibly only to be shot down on appeal when another judge (or judges) makes a decision on a meaning that has to be accepted. There is, as we can see, no such thing as an objective meaning.

Language is an integral and an intrinsic part of humans, and as such is in a state of flux, of change, of chaos, of indeterminate consciousness, of dissonance, of fuzziness. The self is all of those, constantly changing over time, reinventing itself, never stable. How can language be stable when it is part and parcel of humanness?

The very fact that the constantly individually moving *now* is part of all that is perceived by each of us, accounts for why we can define anything only approximately and why we cannot refer to anything as being objectively true. In that respect, feelings, emotions, sentiments and opinions are similar to colors: intrinsically subjective, seen only from one's own perspective and therefore relative to others'.

You cannot feel someone else's pain, however close they are to you and however graphically they describe it. The same applies to taste and smell. How do you describe the sensation from a familiar object, such as the taste of coffee? Or describe the taste of something you have never eaten (say, alligator meat)? How would you do it? 'It's like chicken,' (or 'fishy chicken') is a popular and humorous way to do it, since everyone is presumed to know what chicken tastes like. The buck stops there, though, if the person has never tasted chicken; we can only compare something with what we have experienced when the moving *now* is able to move it back into memory.

If someone claimed to be happy, it would only be understood in relation to what others feel as happiness. Or let's take 'fair'; what could that mean in the sense of 'I bring my children up to be fair'? The word 'fair' is so open to interpretation as to be almost meaningless by any standard. Is it fair for a municipality to charge for water? How much money would be fair? If two nations dispute a piece of land, which is the fair solution?

It is clear, then, that language alone is not adequate. There are no 'objective meanings' that we could simply look up; if there were, notions such as 'consciousness,' 'self,' 'free will,' 'feminism' or, indeed, 'time,' would be understood and discussions would simply be about language.

It was Ludwig Wittgenstein who had famously placed language in that position, claiming in his *Tractatus Logico-Philosphicus* that philosophy boils down to nothing more than a series of linguistic puzzles. It is what people nowadays mean when they stop a discussion with 'define your terms' or triumphantly clinch an argument with 'it's all a matter of semantics.' Unfortunately, it isn't. It can't be. Replacing philosophy with linguistics is not the answer since, as we have noticed, language is approximate, fuzzy and dissonant—like everything else, in fact.

Approximation of language is all-pervasive and keeps a whole industry in business: lawyers, judges, court officials, philosophy and language departments with their critical reasoning classes at universities and colleges, writers and journalists, not to mention politicians, whose every word has to be measured and is thereafter remeasured, dissected and reassessed.

Approximation and simplicity are bedfellows. Approximation necessarily leads away from focus and exactness, so that the greater the approximation, the wider the message. It is the simplicity of President Trump's language that resonates with his audiences and that legitimize their separation from the supposed elite. Umberto Eco had made similar comments about another TV personality, popular Italian game show host Mike Bongiorno, in the early 1960s. For in him, Eco argued, "the spectator sees his own limitations glorified and supported by national authority."[146] Who knows, though? Perhaps a TV personality-turned-politician is an example of how the 18th-century French politician Talleyrand once cynically put it, that language was invented so that we could conceal our thoughts from each other.

On the other hand, why has a universal language, such as Esperanto, not been devised that actually caught on widely? It's not merely a matter of nationalism (although that would have played a part), but most likely its underlying structure as a universal, 'simpler' language with consistent rules in its syntax and morphology without exceptions that stymied its spread. Consistency will always break down in time. Esperanto's draw was that it was 'simple,' yet it didn't manage to stay so.

Language, an integral part of what it is to be human, includes ambiguity and fuzziness. Let's face it: humans are not simple and language must necessarily encompass a way of reflecting the multitudinous aspects of the self.

It so happens that even Wittgenstein changed his opinion about his earlier conclusions, admitting in his *Philosophical Investigations* that perhaps the ideas in *Tractatus* had been oversimplified, and that problems within philosophy need to be solved through looking at how language is actually used, its ambiguity and how it changes. No more immune than events and humans who go through them, a word does not have a fixed definition, but is an evolving entity that carries its own history with it through time, picking up new nuances and discarding old ones through its pragmatic usage, rather than semantic meaning.

All this leads us back to a definition of 'time.' And here we get to a problem even worse than we have encountered till now. Time—of which the moving *now* is the core—can be defined only through what it *does*. It's no wonder that a definition, even by physicists, has not yet been found (or agreed upon). Time is not an 'is.' Time is what it *does*. There is no existence without time. It's not only a matter of the nature of time, but also of what our place is within it; in other words, how it affects us.[147]

For that reason alone, any attempt at a definition of time is doomed to failure. Defining requires being outside that which we attempt to define. We can define a tree, a house, water or any object that is not us. We cannot, though, step out of ourselves in order to define the self or consciousness; we cannot see ourselves other than from where we are as part of ourselves. So while we can define an object (even approximately, enough for it to be understood by others), because the object is outside of ourselves, we have no way of doing that about our inner self. So it is with a definition of time, which is change as part of our self, since we cannot be where we are not. There is no place outside of ourself;

wherever we go, we take ourselves with us. There is no 'view from nowhere,' as the philosopher Thomas Nadel calls it.

Philosopher Huw Price's purpose, in his book *Time's Arrow and Archimedes' Point: New Directions for the Physics of Time*, was to be objective about time. He did say, in parenthesis, that "We can't actually stand outside time, but we can imagine the physics that could." Actually, we can't. However hard we try, we cannot imagine a world that has no time; time would be passing for us even as we design physics for such an imaginable world. We cannot, in other words, stand outside time.

Time is in a class of its own. Other entities—dog, house, stone, air, goodness, space, gravity, desire, art, blue, and so on—exist in relation to its nonexistence (not dog, not house, not stone...). Time is different: time cannot *not* exist. There can never be 'not time,' since there would be no world and there would be no us. Time is the underlying thread/stream of existence.

The eternal *now* ('we are always in the *now*') and the instant *now* ('we want it now') are two sides of the same coin in a culture that accepts intuition at its face value, and that what is perceived is what there is. Often known as the I-generation, it would be more apt to call it the now-generation, that goes for technology, gear, efficiency, training and focused minds in hard bodies. It accounts for the spate of TV and movie superheroes, and zombies where the elite humans can overcome non-fit, mindless half-humans wandering around with nothing to do but feast off the weak 'real' humans. The eternal *now* and the instant *now* reign dissonantly supreme in a world where movie stars within the 'celebrity' genre are taken seriously on any topic, in which their views on world affairs or justice or ethics are given weight for no reason other than they are well known. A belief in that sort of *now* is what brings a TV celebrity to the presidency of the USA, where short un-thought-out answers count for more than experience. It is when *now* is considered to be all there is, that celebrities visiting a foreign country expect to, and do, meet the head of government, where the President of the USA is judged upon which celebrities are invited to the White House and which ones agree to go. In 2014, it was exemplified by the Danish prime minister, Helle Thorning-Schmidt, leaving her official limousine to approach *Sex and the City* actress Sarah Jessica Parker and saying: "I'm the Danish Prime Minister... please can I have your autograph?" Even more amazing is that the crowd were all staring, mesmerized at the actress, not at their prime minister.

Everything?

If time is synonymous with change, and everything changes, should we not assume that everything is time? Is, then, the study of everything the study of time? The big question that follows is whether we can ever be in a situation whereby in understanding time, we understand everything, so that there is nothing left to know and nothing to explain.

The wish is there. "My goal is simple," Stephen Hawking once said. "It is a complete understanding of the universe, why it is as it is and why it exists at all." That, in a nutshell, is the scientific manifesto—ambitious, maybe, but supposedly possible in theory. But it isn't.

On a large scale and which scientists refer to when they talk about 'a theory of everything,' is the aim to reconcile general relativity and quantum mechanics. In *A Brief History of Time,* Steven Hawking famously promised that finding "a complete theory" would allow us to "know the mind of God."

Without thinking about it too much, though, getting to a theory of everything looks promising. After all, scientists in their various fields are finding out more and more about everything they examine. It looks like, as Peter Atkins says in his book *Creation Revisited,* that "We are almost there. Complete knowledge is just within our grasp. Comprehension is moving across the face of the Earth, like the sunrise."

We don't need to go much further to realize that 'complete knowledge' even about one thing is a tall order, so that complete knowledge of everything is daunting to the extreme. Setting aside the hot potato of 'knowledge' (which we have already touched upon), we need to ask if a theory of everything is at all possible. It would have to cover not only what is known, but also take into account what is still to be discovered. Not only that, but a theory of everything is also one that needs to unify all theories in order to account for the basis as to how nature works. Just to add to our woes, the theory itself would have to be valid in the future, as well. But that which is still not known is in the future, which is not part of time and is merely potential for probable knowledge.

Behind the urge for a universal, accepted theory of everything is the implication that there exists an objective real state of nature, that can one day be found. This would mean that there are no aspects of reality intrinsically beyond humans, that all is within the scope of what scientists can conceptually grasp. With that thought, we are well and truly stuck, since we can never know if there is such a barrier to human potential. Up until now, humans have periodically hit problems that were unsolvable at the time; just as quantum theory was beyond the first primates, so our knowledge and therefore our capacity for wider thought will expand in order to leap over the next barrier. With theories more satisfactory than the previous ones, and with none of them being objectively true, it seems unlikely that we would know if a theory of everything has hit the end of the continuum of human knowledge.

If there were such a thing as a theory of everything, it would mean the end of searching (there would be no need), the end of science (everything would be known) and the end of humanness as we know it. That hasn't stopped people from entertaining the thought that there must be something out there that is the ultimate reality waiting to be discovered. But with ultimate reality an impossible dream, science looks for approximations. When David Berman at Queen Mary University of London says that "In some sense, science has always been about approximating reality," we can but wonder if approximate reality is any more realistic than approximate virginity.

There will always be barriers to jump over, to dig under or to knock down. At the root of it all is time. Continuously moving, time is, metaphorically, the engine turning the drum of a cement mixer that keeps its contents from setting. It is because time is everything that we cannot have a theory of everything. There isn't anything that can be in a state of stasis. With everything in movement, and *now* moving to the past even as it reaches toward the future, there is no point at which everything can be captured in a state of everythingness.

There is, though, a danger (and temptation) of seeing the enormity, and therefore the impossibility, of knowing everything as being close to knowing nothing; in other words, coming face to face with a wall of incomprehension, rather than an infinitely stretching future knowledge. Instead of Socrates' "The only thing I know is that I know nothing," and Saint Augustine going a stage further with his "I am in a sorry state, for I do not even know what I do not know," we might be tempted to apply Wittgenstein's "Whereof one cannot speak thereof one must be silent." As a means to shutting up even the slightly curious is Lao Tzu's "Those who know do not talk. Those who talk do not know." Not even allowing an 'if,' this seems to be the ultimate conversation-stopper. But when one gets down to it, the real result of those three different sayings seems to be that we end up listening to those we already agree with. If we are told to keep quiet by someone we believe (or believe in), it is basically a circular invitation to not listen to what someone else is saying.

The desire to know is, and always has been, paramount in

humans. If survival (of the fittest/the most adaptable) means anything regarding humans, it would include the urge to know more than what they already know. For that reason alone, it will never come about that humans know everything. That does not mean that we shouldn't aim toward it, since—and here is another paradox to add to those we have already come across—we will continue even though we know it will never be completed. It is the dissonance of knowing that we won't know everything while aiming toward knowing everything.

This presents a problem for believers in the eternal *now*, for whom *now* is where we are and that we cannot be anywhere else but in the present moment. It's not a matter of *now* being a part of time; we have already seen that it is not. It's the other aspect of *now* that is problematic: its 'mystical' or 'spiritual' element, which is very tempting and widespread because of that. Above all, the idea of *now* as 'what is, is' takes away pressures as to the future and deletes blame about what happened in the past.

Zen (and New Age), more than anything else, is about reclaiming and expanding the present moment. The message is that there is no purpose in getting anywhere if, when you get there, all you do is think about getting to some other future moment. Here is how American spiritual teacher Ram Dass puts it: "As long as you have certain desires about how it ought to be you can't see how it is." As a prescription for the deadening of desire as a way of abandoning free will, it is fairly clear. It certainly goes against the message of 'self-help,' which espouses methods as to what people should do in order to improve their lives. Indian spiritualist Nishan Panwar tells us that "If you live your life being worried about what's going to happen next, then you're going to miss the wonderful things that are right in front of you." Is that advice not to worry about rent, mortgages, tax payments, personal responsibilities to family and friends, etc? Perhaps the oft-quoted expression 'If you aim for nowhere, that's just where you'll go,' is more to the point.

In that *now*, there are no dreams and there is no aiming, at all. Martin Luther King's "I have a dream," on the other hand, has no place in the *now*; it is of the future and action to bring his dream to actuality.

Improvement, needless to say, *does* require you to focus on there being a future. There is no 'go for your dreams, aim high,' which is the hallmark of graduation ceremonies. It can start early. As a headline in *The Telegraph* (UK) screamed: "Pupils in school do not aim for high grades. Teachers 'to blame' for lack of ambition among pupils."[148] It's the politically correct fairness-model encapsulated in the notion that 'the most important thing is taking part, not in winning.' Drubbed the Trophy Kids by writer Ron Alsop, they were brought up to believe that there should be no winners and no losers, all for the sake of harmony as the be-all and end-all. Everyone, no matter how ineptly they perform, gets a trophy.

The urge toward harmony as a leveler can bring its own extremes and a resultant dissonance. Attempting to keep within the so-called comfort zone results in the stance that inviting controversial speakers to speak at universities is to be avoided, since it would harm students by getting them to deal with issues that go against their political principles. Managers complain that as these new graduates moved into the workforce, they expected the workplace to replicate the cozy, non-competitive environment of school, and are shocked to find that the world is not naturally harmonious and that *the present* simply does not last.

The coziness of an everlasting *now* has its roots elsewhere. It is the notion that hanging on to, and concentrating on, *the present* will solve problems of the future. Eckhart Tolle recommends to someone who is worried about paying rent not to think about the future, but only to think about 'here and now.' As a method for relaxing, it sounds good, but it seems unlikely that the landlord will be terribly impressed when the time comes for rent to be paid.[149] This somewhat reckless advice is similar to Joseph Campbell's: "A bit of advice given to a young Native American at the time of his initiation: As you go the way of life, you will see a great chasm. Jump. It is not as wide as you think." Yes, but often it is wider than you think!

The self and identity

In normal circumstances, we don't need to think about others' identity. We recognize people and simply know who they are. Even if we haven't seen them for a while and consider that they 'have changed,' we still think of them as the same person. That applies even if they have changed radically physically and/or mentally, say after a stroke. And despite the knowledge that we are not simply our bodies—we accept the scenario in novels and movies about people exchanging bodies—we look to the physical body to tell us of someone's identity or self, and accept that it is one's DNA that makes us 'us' and each person 'them.'

Despite that, we are aware that each of us is relentlessly changing. We know and feel that we aren't the same as when we were children or when we were teenagers and we are also well aware that we will feel differently about today when we are in our seventies. Yet, we also feel that we continue to possess our essence, a core, that we are 'me' all the time, the fine balance—the dissonant balance mentioned earlier—of our oscillating consciousness, that which Julian Baggini calls a sense of 'me-ness.' But where is the 'me' at any point in time? After all, we look at the world from the viewpoint of the currently shifting 'me' that is taking place *now*. It is never 'me,' because no sooner is it 'me' than it has already become a different 'me.' It is not surprising, then, that there are different selves within the same body: parent, sibling, child, friend, employee, employer… who come out when appropriate.

This puts an interesting slant on relationships. How does love work? How can we love someone who is not one self, but is constantly changing? Which part of that self do we love in those circumstances? Perhaps, in fact, we love who we want the person to be. Since there is no static self, do we love the idea of being in love, rather than having an object to our love? These are questions that may not have definitive answers, because they relate to a continual changing self.

Questions, though, may bring out further questions and perhaps shake ensconced ideas. Take, for example, the widespread notion of the necessity of 'finding oneself,' which is synonymous with the oft-used 'finding one's true self.' But from what we have seen, the notion of a 'true self' hidden deep down, waiting to be discovered, is a dubious proposition. Catching hold of a constantly moving self is akin to holding onto a greased pole. And, yet, there is an obsessive interest in looking for and finding that true self through self-awareness, spurred on by a whole industry of self-seekers.

Interestingly, but not surprisingly, many dedicated searchers find nothing specific there. In her book,[150] spiritualist teacher Suzanne Segal related her experiences of losing her 'self' completely, so that when she is asked who she is, she says that she is an infinity, a 'vastness.' "I am no one and everyone, nothing and everything, just as you are." The Catholic New Ager/post-Christian mystic Bernadette Rogers found what she called "the path to no-self," set out in her book *The Path to No-Self: Life at the Center*, and three years later in another, with a similarly oxymoronic title, *The Experience of No-Self: A Contemplative Journey*. Despite being a Catholic nun, she had become enamored of Buddhism, claiming that it held "a profound and difficult truth," thereafter leaving her convent, marrying and having four children. It was a difficult path, she maintained. From nun-self to non-self.

The idea of 'no-self' is a cardinal point within New Age, since it reflects the eradication and annihilation of anything that could come in the way of being part of the wholeness of the universe. It is where the 'true self' is accepted as an egoless 'no-self.' Instead of "wherever you go, there you are," it would perhaps have been more appropriate for the American Buddhist nun Pema Chodron, who we met earlier, to have said, "wherever you go, there you are not."

Presumably, the aim of those searching for their no-self is to have a better and more authentic life, while being spared from day-to-day irrelevancies. When actor and singer Harry Dean Stanton was asked how he described himself, he replied, "As nothing. There is no self." He then went on to explain that he was big into Eastern concepts. "The horror of life, the love of children, the whole phantasmagoria—it's all meaningless." His solution? "Be still, and see what happens. All of life unfolds perfectly." Destiny and fate, it seems, is his answer. "You have to get beyond consciousness," he concludes. What that entails is not made clear, but when asked what he does, his answer was that he watches television game shows.[151]

All of that puts the cat among the pigeons when it comes to time. If one is to aspire to no-self—the ultimate aim of meditation, perhaps the most well-known New Age practice—where would one be when one reaches it? If everything is *now*, as they tell us, where is this *now* when we are in our non-self? Bernadette Roberts has the answer: "By the time the journey [into no-self] is over, the only possible way of living is in the now-moment, wherein the mind moves neither backward nor forward but remains fixed and fully concentrated in the present." Yet, if there is still a mind that does the knowing, in fact is concentrating on the now-moment, it would mean that there is still a self.

For someone's no-self to be fixed, it would require for there to be no change. This, of course, is an impossible state within time. Yet, with the aim of New Age to get to no-self ('the real self') through, for example, meditation, there needs to be constant change in order to reach a more enlightened plane.

In a review of *Mindfulness for Health: A practical guide to relieving pain, reducing stress and restoring wellbeing,* Professor Mark Williams refers to the 'real self,' when he states that the book "will put you back in touch with the extraordinary person you already are." That leaves us with a facetious question: What if the extraordinary person we find inside is not that great after all, not a paragon of goodness that we expect to be hidden in there? The assumption is always that what we find we will want to embrace with alacrity. It is a valid assumption only because it is our subjective self that is searching for an objective 'real' self. It goes without saying—or it should do!—that because one can see only subjectively what is really happening inside oneself, one would not be able to find one's real self even if there were such a thing. There is nowhere one can stand in order to do that. Just like parents cannot be objective about their own children, however much they are convinced that they are, we are even less likely to be objective about ourselves.

But in the age of speed and instant everything, there are, it seems, simpler and less painful ways of reaching awareness, as actress Emma Watson discovered. "I think it's amazing," she says, "how self-aware people are becoming as a result of constantly posting images on Facebook and Instagram." Whether messaging and reacting on social media as acts of self-propagation help in attaining self-awareness is interesting speculation.

In an era that expects everything to be *now,* where *carpe diem* rules and *now* is all that matters, it is perhaps not surprising that there has been an increasing rate of narcissism, as testified by sociological studies.[152] The reason given is that the increase in narcissism (a 'positive and inflated view of the self') is related to the fact that Western culture has increasingly celebrated individualism.

In itself, this is not unique. Every generation has been the most narcissistic ever. Here's a quote from an *Atlantic* article in 1907: "The rock upon which most of the flower-bedecked marriage barges go to pieces is the latter-day cult of individualism; the worship of the brazen calf of the Self." But while young people have always been narcissistic, in this particular I-generation it is touted loudly as something to be proud of, to aim for. A whole industry has been built around it. When self-help is about how to make the individual the center, and when often the largest section in bookstores is the seemingly endless array of self-help books, it says something. In previous times, narcissism was the position of young people, while today it is for all ages. If in the past young people got over it as they grew older, nowadays self-centeredness is a way of life for all ages. Nobody is too old to benefit from meditation, mindfulness exercises, and the countless seminars and courses that are given by a mélange of gurus willing to pass on the secret of personal happiness.

The concept of *now* as a timeless period into which one delves in order to find one's self plays an integral part in the act of isolation. It is not surprising that the cosmetic company L'Oréal's ad campaign "Because you're worth it" was so successful. It should also come as no surprise that the trend of self-gifting, which started as a treat for Christmas but is spreading beyond, is on the rise, not to mention the trendy sologamy: marriage to oneself, replete with a self-marriage ceremony, which includes vows to oneself as commitment to self-love and self-care.

Why

One thing is clear: it's not always easy to differentiate between the present, past and future. Weaving through change, which often mirrors itself onto different aspects of perspective that are in constant motion, time is difficult to pin down—literally. Here's an example to illuminate the problem. Take traffic jams, which can form out of the simplest of situations. One driver gets too close to the car in front and has to brake, as does the driver behind, as does the driver behind him, causing a stop-and-go shock wave down the highway. Watching this unfold from the side of the road, a bystander sees the cars behind as the future of the cars in front of them before each is hit. But for the drivers in the cars in front, the cars behind them are in the past on the road already taken. Yes, but... the car in front is causing all the other cars to slow down, so that they are in the future along that road already taken. Mind-numbing. But that's time.

Let's look at the scene again, using our moving *now* to clarify what is happening. The shock wave caused by the first car is, graphically and metaphorically, the backward-moving *now*. When the first driver presses the brakes, *now* is instantly a memory, moving down the line from driver to driver. Each driver is slowing down because of what has happened in front and they are aware of what they think will happen in the future. Their predictions will remain as such as the moving *now* becomes a memory and the future looms as certain. When it happens, the event will be the past and only then can they discern if the prediction panned out. The prediction may prove to have been wrong; for example, if they didn't actually crash into the car in front of them. The question as to whether the car behind was in the past or in the future does not come up when the moving *now* is taken into account: both the car in front and the car behind are in the past.

We have been talking about time as if it goes through the present to the past (or from the future through the present to the past). But we could equally claim that it comes to us from the past. We can think of it as moving events, somewhat like waves from the future that meet those from the past, to create what we perceive as the present. If we add to this the known future (that it will end), we have a situation somewhat akin to theories in quantum mechanics, where the future affects the past and the past affects the future.

The waves of time that break over each other seem to obviate the possibility of harmony, instead showing an underlying movement of dissonance.

Dissonance, fragmentation, and approximation perhaps best sum up the effect that time has on us and the world. They encompass the cause and effect of the ultimate paradox: the move forward to the future that is continuously moving back to the past. The linear movement of time, inevitable and relentless, is inseparately juxtaposed against the inexorable backward-moving *now* within the symbiotic nature of the past and the future. They have much in common, the past and the future. We know there was a past but do not know what it was. Likewise with the future, we know that there is a future but we don't know what it will bring.

Time is itself a paradox, a moving dissonance within itself, fragmented and approximate. The universe, and everything in it, is in constant flux. This constant movement is the nature of the universe, with everything relative to everything else, non-static by their very nature.

A point made constantly here is that things may not be what they seem. But that's a gentle way of putting it. Things are *never* what they seem. They can't be. To be blunt, we don't know 'how things are'; we can only know how we think they are. In that respect, with everything in constant movement (and thus changing), there isn't anything that can be pinned down as 'the way it is.'

As much as that solves problems and answers questions, it can also hinder us in other ways. We cannot, for example, attempt answering the question of questions: What is the secret of life? We can't, because nothing is ever in a fixed position on which to latch the question. For the same reason, we cannot describe the nature of reality—a favorite philosophical question. In fact, that could not be our aim. With the continuously backward-moving *now*, there can be no reality, no objective truth 'out there.' It matters not if a consensus points to a supposed definite Truth. A thousand or a million people, each with a different *here* and *now*, cannot point to idealistic Truth, be it in science, art, music or philosophy.

Time unites as well as separating us from each other. Within our interactions in the world, while we are each in our own backward-moving *now* and in our own *here*, we nevertheless live together in the common perception of *the present*. As individuals, though, nobody else is *here* and nobody else is *now*.

Is it important to know that? In the scheme of things, perhaps not. It is, though, an explanation as to why things don't seem to go as we always expect them to, why sometimes nothing seems to be steady, why the ground itself seems to be moving under us.

Unlike most religions, faiths, beliefs and all self-help programs, whose purpose is to make people content, the job of philosophy is to question that contentment, to get people to face what there might be. It is about understanding that whatever we find to be obvious (or 'the Truth'—with or without a capital T) may eventually and even inevitably be seen to be merely a marker toward another 'obvious truth' in the future, or perhaps even to an earlier one in the past. While religions expect gratitude, philosophy wouldn't mind receiving it but would carry on regardless. "Why do people accept that things are the way they are when they could be some other way?" asks Argentinian novelist Julio Cortázar.

Dismissing 'now' as a period in time is a change in thinking. It is a change of direction, a different way of looking at time and thus at life. Or as a line for the movie *Ferris Bueller's Day Off* puts it: "Life moves pretty fast, if you don't stop to look around every once in a while, you might miss it."

When Bishop Berkeley made the observation that "Philosophers have raised a dust and then complain that they cannot see," it may often have been valid. Our walk through time, though, has raised a dust in order to illuminate the movement at its heart.

By subjecting every belief to critical reflection, we move along a path of inquiry on a continuum of approximate understanding. Always on the verge of the next, we are constantly on the way to possible better decisions that lead to independent thought and abandonment of the deceptive stories we tell ourselves. That doesn't mean that our independent thought will lead us to the truth; we shouldn't expect that. We can't expect that. But an understanding that satisfies us will help us to hold onto the continuously moving time.

Now that we have come to the end of this particular journey, we can ask: What difference does it make to us if we accept *now* in a different way than we did before? 'What's the point?' is a question that has often been asked over the ages about new discoveries, such as when Socrates asked what 'good' it was to know that the Earth was round. We asked that question about *now* on the first page of this book. After all, it won't make a difference in our daily lives; we will continue to live in *the present* and accept *now* as we did before. But it does make a difference. After our rummage into time and *now*, you might be looking at things differently. That the world contains physical distinct objects has not been questioned. The moving *now* does not mean that trees, animals, buildings and paintings do not exist. It is how we each see them that is questioned. And because *how* and *what* have everything to do with *when*, you may possibly not be able to use the word 'now' without thinking of how it may not actually exist in the way everyone else feels it does. Even if you were not totally (or not at all) convinced by the arguments, you cannot help wondering…

> "*… C'est le temps qui contient l'univers et non le contraire.*"
> ("It is time that contains the universe, not the opposite.")[153]

Notes

1. Available at: http://theinvisiblegorilla.com/gorilla_experiment.html.
2. *Free Will*, 2012.
3. 2004.
4. *The Magic Mountain*, 1928.
5. Available at: https://www.youtube.com/watch?v=78dOAeSV29E.
6. Alexandra Horowitz, 2016.
7. 1912.
8. *Now: Embracing the Present Moment*, 2011.
9. 1990. But it was used earlier, in Ray Cummings' 1922 science fiction novel *The Girl in the Golden Atom*.
10. Steven Taylor, 2008.
11. Nietzsche, 1882.
12. *The Secret*, 2006.
13. *The End of Time*.
14. Available at: http://www.webperformancetoday.com/2013/03/01/how-does-the-human-brain-keep-track-of-time-video/.
15. Available at: https://www.sciencedaily.com releases/2013/08/130822090326.htm.
16. "A matter of Time" at the World Science Festival, 2014.
17. Available at: https://www.youtube.com/watch?v=Oswx23Tz-_w.
18. Paul McCartney, John Lennon Copyright: Sony/ATV Tunes LLC.
19. 2013, Studio Hamburg Filmproduktion.
20. By Sumangali Morhall, Mantra Books, 2012.
21. But it is not logic or rationality that takes a part here, and certainly not when it comes to Eastern thought, which discounts rationality and logic as a means of tackling its concepts. An extreme example is Tao, which tells us that

thinking we understand its precepts means that we don't.

22. Ecclesiastes 3:1–8.
23. *Psychological Science*, July 2013.
24. Granta, 2012.
25. Research by David Eagleman, known for his work on time perception, has shown that time does not actually run in slow motion but, rather, is only a retrospective assessment by the person. We will look at this in the following chapter.
26. *Seinfeld: The Old Man*, No. 4.18, 1993.
27. *The Illusion of the End*, Polity, 1994.
28. 1984.
29. New Age Beliefs and Spirituality forum available at: www.essentialbaby.au.com.
30. Andrew Lutts, Salem New Age Center.
31. It was first popularly used in the motion picture *The Adventures of Buckaroo Banzai Across the 8th Dimension* in 1984.
32. From *Ode Magazine* (by Tijn Touber), Jan/Feb 2006.
33. "A Celebration of Reason," 2012 Global Atheist Convention, Melbourne, Australia.
34. *Nature Neuroscience*, March 30, 2014.
35. TED, February 2012.
36. *Live science*, December 20, 2012.
37. Nicholas Epley and Erin Whitchurch: "Mirror, Mirror on the Wall: Enhancement in Self-Recognition," *The Personality and Social Psychology Bulletin*, June 11, 2008.
38. The American physicist Hugh Everett did not take changing individual memories into account when he stated: "...one can imagine an intelligent amoeba with a good memory. As time progresses the amoeba is constantly splitting, each time the resulting amoebas having the same memories as the parent."
39. "I am not a story," from *On-Life Writing* by Zachary Leader, 2015.

40. *Social Acceleration: A New Theory of Modernity.*
41. *The Emptiness of Existence.*
42. A term made famous by Marcel Proust in his monumental novel *In search of time lost*, as opposed to 'voluntary memory.'
43. Sonnet XXX.
44. Andy Warhol, 1977.
45. *Speak, memory*, 2013.
46. The testimony was taken on April 7–8, 1964 at the Old Civil Courts Building, Royal and Conti Streets, New Orleans, LA, by Mr Wesley J. Liebeler, assistant counsel of the President's Commission.
47. Rudolf, 1999.
48. *Love's Executioner: and Other Tales of Psychotherapy*, 1989.
49. 1989, p. 172.
50. Quoted in Rudolf, 1999.
51. *Moving Violations: war zones, wheel chairs, and declarations of independence*, 1995.
52. "The Deceptive Watchman: How Our Brains Twist Time." Available at: https://www.youtube.com/watch?v=rAKtcxhs4lc.
53. *Psychonomic Bulletin & Review*, September 2002, Volume 9, Issue 3.
54. *The New Yorker*, May 19, 2014.
55. *Scientific American*, September 1997, vol. 277, No. 3.
56. *New Republic*, February 5, 2015.
57. Karl Sabbagh, 2010.
58. Richard J. McNally, 2005.
59. 2010.
60. Discovery.com, February 9, 2015.
61. *Psychological Science*, September 4, 2015.
62. *The Heretics: Adventures with the Enemies of Science*, 2013.
63. *Sybil Exposed*, Debbie Nathan.
64. Durand, Marcella. *BOMB Magazine*. Summer 2009. Retrieved July 26, 2011.

65. Skeptoid, December 23, 2014. Available at: http://skeptoid.com/episodes/4446 [accessed August 29, 2017].
66. February 10, 2015.
67. "On Exactitude in Science," in *Collected Fiction.*
68. Available at: http://www.newscientist.com/article/mg21528781.500-they-never-forget-the-strange-gift-of-perfect-memory.html?page=1.
69. Samuel Arbesman, 2012. *The Half-Life of Facts: why everything we know has an expiration date.*
70. Jacob Browski, from *The ascent of Man,* BBC, 1973.
71. With thanks to N.S.
72. Wikipedia.
73. *The Free Dictionary.*
74. Faro, Leclerc & Hastie, 2005.
75. *The Fog of Peace,* Rifkind and Giandomenico, 2014.
76. Interview in *Chicago Tribune,* May 25, 1916.
77. *Himmler's Crusade,* Christopher Hale.
78. Jude England, head of social sciences at the British Library.
79. *The Invention of Scotland,* 2009.
80. *The Evolution Of God,* Robert Wright.
81. Related graphically by Tony Horwitz in *Confederates in the Attic.*
82. *Ardennes 1944: Hitler's Last Gamble,* Antony Beevor.
83. "Argentina's coup: social myth, memory and history," Federico Guillermo Lorenz, *History Today,* Vol. 54 Iss. 1 January 2004.
84. Attributed to Socrates by Plato, according to William L. Patty and Louise S. Johnson, *Personality and Adjustment,* 1953, although research indicates that the quotation is spurious.
85. *Hancock's Half Hour,* 1959, BBC.
86. *Nature,* January 21, 2016.
87. *Eavesdropping on Jane Austen's England,* Roy & Lesley Adkins.
88. *Confederates in the Attic.*

89. *London Fog: The Biography*, Christine Corton.
90. *The Whig Interpretation of History*, 1931.
91. *Jerusalem: The Biography*.
92. Rabbi Shemuel ben Nachmani, as quoted in the Talmudic tractate Berakhot (55b., paraphrased later by H. M. Tomlinson as: "We don't see memories as they are, but as we are."
93. *Tablet*, February 25, 2015.
94. Guldi and Armitage, 2014.
95. Green, 2011.
96. *Time and Free Will*, 1910.
97. Available at: https://www.youtube.com/watch?v=6yMiUq7W_xI.
98. "The Time We Thought We knew," *The New York Times*, January 1, 2004.
99. Available at: https://www.youtube.com/watch?v=-O8lBIcHre0.
100. The paradox is 'solved' through the concept of the multiverse, the hypothetical set of infinite universes, through which the act of killing one's grandfather in another universe would not affect what was happening in the universe where one is time traveling.
101. *A matter of Time* at the World Science Festival, 2014.
102. D. L. Mahler, T. Ingram, L. J. Revell and J. B. Losos. "Exceptional Convergence on the Macroevolutionary Landscape in Island Lizard Radiations." *Science*, 2013; 341 (6143).
103. *Philosophie Zoologique*, 1809.
104. Robert James Waller, 1992.
105. Carroll 1872/1982, 171.
106. Nicholas Carr, *The Shallows*, 2010.
107. Aminatta Forna, 2014.
108. *The Age of Absurdity: Why Modern Life Makes It Hard to Be Happy*, 2011.

109. *The Religion of the Future*, 2016.
110. Scientists used brain tests on 3 year olds to determine which children are more likely to grow up to become criminals. *Nature Human Behaviour*, December 12, 2016.
111. Directed by Terry Gilliam, 1995.
112. Roger Penrose devised such a situation on a chess board, which should be solvable for humans but would defeat an AI computer. The chess problem—originally drawn up by Sir Roger—has been devised to defeat an artificially intelligent (AI) computer but be solvable for humans. Available at: www.chessbase.in/news/chess-position-by-roger-penrose/#sthash.utUYQ5DQ.dpuf.
113. Johann Wofgang von Goethe, *Goethe, the story of a man: Being the Life of Johann Wofgang von Goethe as Told in His Own Words and the Words of His Contemporaries*, ed. and trans. Ludwig Lewisohn (New York: Farrar, Straus, 1949) p. 353.
114. *Conversations of Goethe with Johann Peter Eckermann*, trans. J. K. Moorhead (Cambridge, MA: Da Capo Press, 1988) p. 287.
115. "The Barbed Gift of Leisure, *The Chronicle of Higher Education*, 2013.
116. *Grief is the Thing with Feathers*, 2015.
117. *Journal of Sport and Exercise Psychology*, August 22, 2016.
118. *Medical News Today*, November 1, 2016.
119. *Essentialism*, 2014.
120. Sometimes there is an exact number that decides an election as, for example, the number of electoral votes needed to decide the USA elections (270). But the number of votes needed to arrive at the majority of electoral votes in each state is not known until the votes are counted.
121. The notion of a universe and time beginning from nothing I discuss elsewhere, in *Nothing Matters—a book about nothing*.
122. "My Writing Day," *The Guardian*, August 6, 2016.
123. This is not the first time in the book that we have suggested

that the future could affect the past. If this *now*-driven retrocausal model has any bearing on similar phenomena in quantum physics, it would be an interesting development. The notion that future events could affect the present and the past is not new. Physicists John Wheeler, Richard Feynman, Dennis Sciama, Yakir Aharonov and today, Huw Price, speculate that causality could run backward. The difference in approach of the moving *now* theory of time is the claim that the future emanates from the past. There is no future already set, a destiny in place. The past and future affect each other without an intermediary *present*.

124. Based on James Bradley's book, *Flyboys*, 2003.
125. *Every Day*, 2013.
126. *The Improbability Principle*, 2014.
127. *The Telegraph*, August 24, 2012.
128. *Manifesting Divine Consciousness in Daily Life*, 2014.
129. Quantum physics, from a different perspective, points to something similar by illustrating the uncertainty of the location of fundamental particles.
130. *Freedom Regained: The Possibility of Free Will*, 2015.
131. *Sapiens: A Brief History of Humankind*.
132. "Time and Impermanence in Middle Way Buddhism and Modern Physics." Talk at the Physics and Tibetan Buddhism Conference University of California, Santa Barbara January 30–31, 1998.
133. Steven Poole, *Rethink: The Surprising History of New Ideas*.
134. Kurt Gödel, for example, did not accept *now* as an *objective* time period, envisaging instead each observer as having their own set of 'nows.' While closer to my position, he nevertheless considered 'now' as a real part of time.
135. *Reality is Not What it Seems: the journey to quantum gravity*, 2016.
136. *Stephen Hawking and the theory of everything*, Acorn Media Group, 2007.

137. Amnesty Lecture, Oxford, February 18, 1997.
138. "How our contradictions make us human and inspire creativity," *Aeon*, December 7, 2016.
139. Following Jean Baudrillard, who stated that as soon as something becomes perfect, it disappears into the realms of the theoretical.
140. "Anthem" from the album *The Future*, 1992.
141. *Scientific American*, May 1963.
142. *The Observer*, July 7, 2013.
143. *The Silence of Animals*, 2013.
144. Although attributed by her to St Francis of Assisi, this was, in fact, composed by a 19th-century follower.
145. On Conan O'brien's Words of Wisdom, Howard Stern Interview 2013.
146. *Fenomenologia di Mike Bongiorno*, 1961.
147. With thanks to Neil Spungin for this discussion.
148. *The Telegraph*, October 26, 2012 .
149. From *Ode Magazine* (by Tijn Touber), Jan/Feb 2006.
150. *Collision with the Infinite*, 1996.
151. Interview on *The New Yorker*, September 30, 2013.
152. Lisa Wade, 2013.
153. Marc Levy, *La Prochaine Fois*, p. 36.

Bibliography

Ades, Dawn. 1982. *Dalí* (London: Thames and Hudson).

Adkins, Roy and Lesley. 2013. *Eavesdropping on Jane Austen's England* (London: Little, Brown Book Group).

Arbesman, Samuel. 2013. *The Half-Life of Facts: why everything we know has an expiration date.* (New York: Current The Penguin Group).

Atkins, Peter. 1993. *Creation Revisited* (W.H. Freeman & Co Ltd).

Baggini, Julian, 2015. *Freedom Regained: The Possibility of Free Will.* (Granta Books).

Ball, Philip. 2014. "Beauty is Truth?" *Aeon Magazine.* May 19, 2014.

Barbour, Julian. 2011. *The End of Time* (Oxford University Press)

Baudrillard, Jean. 1992. *The Illusion of the End* (Stanford, Ca: Stanford University Press).

Bayley, Stephen and Mavit Roger. 2011. *Life's a Pitch: How to Sell Yourself and Your Brilliant Ideas* (London: Transworld Publishers).

Beevor, Antony. 2015. *Ardennes 1944: Hitler's Last Gamble* (New York: Viking).

Bergson, Henri. 1910. *Time and Free Will.* Trans. F.L. Pogson (London: George, Allen & Unwin).

Birkerts, Sven. 2015. *Changing the Subject: Art and Attention in the Internet Age* (Minneapolis, MN: Graywolf Press).

Boorstein, Sylvia. 1998. *That's Funny, You Don't Look Buddhist: On Being a Faithful Jew and a Passionate Buddhist* (New York: HarperOne).

Borges, Jorge Luis. 1998. "On Exactitude in Science." *Collected Fictions*. Trans. Andrew Hurley (New York: Penguin Putnam).

Bower, Tom. 2012. *Sweet Revenge: The Intimate Life of Simon Cowell* (New York: Faber and Faber).

Bradley, James. 2003. *Flyboys: a true story of courage* (New York: Little, Brown Book Group).

Burch, Vydiyama and Penman, Danny. 2013. *Mindfulness for Health: A practical guide to relieving pain, reducing stress and restoring wellbeing* (London: Piatkus).

Butterfield, Herbert. 1965. *The Whig Interpretation of History* (New York: Norton).

Byrne, Rhonda. 2006. *The Secret* (London: Atria Books/Beyond Words).

Carr, Nicholas, 2010. *The Shallows: How the internet is changing the way we think, read and remember* (New York: Norton).

Carroll, Lewis. 2011. (1865) *Alice in Wonderland* (New York: Empire Books).

Chabris, Christopher and Simons, Daniel. 2011. *The Invisible Gorilla* (New York: Crown Publishers).

Chopra, Deepak. 1994. *Ageless Body, Timeless Mind: The Quantum Alternative to Growing Old* (New York: Three Rivers Press).

Choquette, Sonia. 2013. *Tune In: Let Your Intuition Guide You to Fulfillment and Flow* (New York: Hay House).

Clune, Michael. 2013. "The Quest for Permanent Novelty." *The Chronicle Review*, February 11, 2013.

Corton, Christine. 2015. *London Fog: The Biography* (Cambridge, MA: Harvard University Press).

Cummings, Ray. 1922. *The Girl in the Golden Atom*. Republished 2009 by CruGuru (online).

Dass, Ram. 1971. *Be Here Now* (Taos, NM: Lama Foundation).

Dawkins, Richard. 2014 (12 January). "Essentialism." *The Guardian*, UK.

Dunning, B. "Ganzfeld Experiments." *Skeptoid Podcast.* Skeptoid Media Inc. Web. Feb 5, 2013. http://skeptoid.com/episodes/4348.

Durand, Marcella. 2009. "Jacques Roubaud." *BOMB Magazine*.

Durkheim, Emile. 1912. *The Elementary Forms of the Religious Life*. Trans. Karen E. Fields. (Oxford: Oxford University Press, 2001).

Eagleman, David M. 2009. Edge.org. http://edge.org/convers

ation/brain-time.

Einstein, Albert. 1920. *Relativity, the special and the general theory* (London: Methuen).

Faro, D., Leclerc, F. and Hastie, R. 2005. "Perceived causality as a cue to temporal distance." *Psychological Science* 16, pp. 673–677.

Flintoff, John-Paul. 2012. *How to Change the World* (New York: Macmillan).

Foley, Michael. 2011. *The Age of Absurdity: Why Modern Life makes it Hard to Be Happy* (London: Simon & Schuster UK).

Forna, Aminatta. 2013. *The Hired Man* (New York: Atlantic Monthly Press).

Foucault, Michel. 1984. "Of Other Spaces, Heterotopias." *Architecture, Mouvement, Continuité* 5.

Freitas, Donna. 2013. *The End of Sex: How Hookup Culture is Leaving a Generation Unhappy, Sexually Unfulfilled, and Confused About Intimacy* (New York: Basic Books).

Gelernter, David. 2013. "The End of the Web, Search, and Computer as We Know It." http://www.guardian.co.uk/technology/2013/feb/10/stream-consciousness-computer-desktop.

Gladwell, Malcolm. 2007. *Blink: The Power of Thinking without Thinking* (New York: Back Bay Books).

Gleiser, Marcelo. 2010. *A Tear at the Edge of Creation: A Radical New Vision for Life in an Imperfect Universe* (New York: Free Press).

Golding, William. 1959. *Free Fall* (New York: Harcourt).

Gray, John. 2013. *The Silence of Animals: On Progress and Other Myths* (New York: Farrar, Straus and Giroux).

Green, Ronald. 2011. *Nothing Matters—a book about nothing* (Airesford, Hants: iff-Books).

Grudin, Robert. 1982. *Time and the Art of Living* (Cambridge, Mass: Harper & Row).

Guldi, Jo and Armitage, David. 2014. *The History Manifesto*

(Cambridge: Cambridge University Press).

Hale, Christopher. 2013. *Himmler's Crusade: The Nazi Expedition to Find the Origins of the Aryan Race* (Hoboken, NJ: John Wiley & Sons).

Hand, David J. 2014. *The Improbability Principle: Why Coincidences, miracles, and rare events happen every day. Scientific American* (New York: Farrar, Straus and Giroux).

Harari, Yuval, Noah. 2016. *Homo Deus: A Brief History of Tomorrow* (London: Harvill Secker).

Harris, Sam. 2012. *Free Will* (NY: Free Press).

Harvey, Jason. 2010. *Achieve Anything in Just One Year: be inspired to live your dreams and accomplish your goals* (Canada: Amazing Life Press).

Hawking, Stephen. 1998. *A Brief History of Time: The Updated and Expanded Tenth Anniversary Edition* (New York: Bantam Books).

Hertz, Noreena. 2006. *The Debt Threat: How Debt Is Destroying the Developing World... and Threatening Us All* (New York: HarperBusiness).

Hockenberry, John. 1995. *Moving Violations: war zones, wheel chairs, and declarations of independence* (New York: Hyperion).

Holt, Jim, 2013. *Why Does the World Exist?: One Man's Quest for the Big Answer* (London: Profile Books).

Horowitz, Alexandra, 2016. *Being a Dog: Following the Dog Into a World of Smell* (New York: Scribner).

Horwitz, Tony. 1998. *Confederates in the Attic* (New York: Vintage).

Joyce, Rachel. 2012. *The Unlikely Pilgrimage of Harold Fry* (London: Doubleday).

Kabat-Zin, Jon. 2004. *Wherever You Go, There You Are: Mindfulness meditation for everyday life* (London: Piatkus).

Kahneman, Daniel. 2011. *Thinking, Fast and Slow* (New York: Farrar, Straus and Giroux).

Kingwell, Mark. 2013. "The Barbed Gift of Leisure." *The Chronicle of Higher Education*. March 25.

Krishnamurti J. and Bohm, D. 1985. *The Ending of Time* (New York: Harper & Row).

Kurzban, Robert. 2012. *Why Everyone Else Is a Hypocrite* (Princeton, NJ: Princeton University Press).

Langer, Ellen. 1989. *Mindfulness* (Boston, Mass: Da Capo Press).

Leader, Zachary (ed.). 2015. *On-Life Writing* (Oxford: Oxford University Press).

Levithan, David. 2013. *Every Day* (New York: Ember).

Lipovetsky, Gilles. 2002. *The Empire of Fashion: Dressing Modern Democracy* (Princeton University Press).

----- 2005. *Hypermodern Times* (Cambridge, UK: Polity).

Litt, Toby. 2012. The Reader and Technology. *Granta,* April 12.

MacKenzie, C. and Atkins, K. (eds). 2008. *Practical Identity and Narrative Agency* (NY & London: Routledge).

Mann, Thomas. 1924. Trans. Martin Secker 1928. *The Magic Mountain* (New York: Penguin Books, 1960).

Maran, Meredith. 2010. *My Lie: A True Story of False Memory* (San Francisco, CA: Jossey-Bass).

Marcer, Peter J. 1995. "The need to Define Consciousness—A Quantum Mechanical Model," pp. 3–15, *Consciousness: The Big Issue,* Symposium, ed. A. M. Fedorec & P. J. Marcer (Greenwich, UK: University of Greenwich).

Markides, Kyriacos C. 2012. *Inner River: A Pilgrimage to the Heart of Christian Spirituality* (New York: Random House).

Markosian, Ned. 1993. "How fast does Time Pass?" *Philosophy and Phenomenological Research* 53 (1993), pp. 829–844. Reprinted in William R. Carter (ed.), *The Way Things Are: Basic Readings in Metaphysics* (New York: McGraw-Hill, 1998) pp. 129–147.

Mata, Sri Mrinalini. 2014. *Manifesting Divine Consciousness in Daily Life.* (Los Angeles, CA: Self-Realization Fellowship).

Mayer-Schönberger, Viktor. 2011. *Delete: The Virtue of Forgetting in the Digital Age* (Princeton, NJ: Princeton University Press).

McCarthy, Corma. 1993. *All the Pretty Horses* (New York: Vintage).

McNally, Richard J. 2003. *Remembering Trauma* (Cambridge, MA:

Belknap Press).

Montefiore, Simon Sebag. 2011. *Jerusalem: The Biography* (New York: Alfred A. Knopf).

Morhall, Sumangali. 2012. *Western Disillusionment to Eastern Spiritual Fulfilment* (Airesford, Hants: Mantra Books).

Nagel, Thomas, 2012. *Mind and Cosmos: Why the Materialist Neo-Dawinian Conception of Nature is Almost Certainly False* (Oxford: Oxford University Press).

Nathan, Debbie. 2011. *Sybil Exposed* (New York: Free Press).

Neill, Michael. 2013. *The Inside-Out Revolution: The Only Thing You Need to Know to Change Your Life Forever* (New York: Hay House).

Nietzsche, Friedrich. 1882. *Die Fröhliche Wissenshaft.* Trans. (2001) as *The Gay Science: With a Prelude in German Rhymes and an Appendix of Songs*. Ed. B.A.O. Williams, trans. J. Nauckhoff, poems trans. A. Del Caro (Cambridge: Cambridge University Press).

Nora, Pierre. 1984. *Les Lieux de mémoire 1* (Paris: Gallimard).

O'Connell, John. 2012. *For the Love of Letters* (London: Short Books).

Penfield, Wilder. 1984. The *Mystery of the Mind* (Princeton: Princeton University Press).

Penrose, Roger. 1989. *The Emperor's New Mind: Concerning Computers, Minds, and the Laws of Physics.* (Oxford: Oxford University Press).

Poincaré, H. 1902. *Science and Hypothesis* (London: Walter Scott Publishing Co.).

Poole, Steven. 2006. *Rethink: The Surprising History of New Ideas* (New York: Random House).

Porter, Max. 2015. *Grief is the Thing with Feathers* (New York: Faber).

Price, Huw. 1996. *Time's Arrow and Archimedes' Point: New Directions for the Physics of Time* (Oxford: Oxford University Press).

Rifkind, Gabrielle and Picco Giandomenico. 2014. *The Fog of Peace* (London: I.B. Taurus).

Roberts, Bernadette. 1993 (revised edition). *The Experience of No-Self: A Contemplative Journey* (New York: State University of New York Press).

--- 1985. *The Path to No-Self: Life at the Center* (Boulder, CO: Shambhala Publications).

Rosa, Harmut. 2013. *Social Acceleration: A New Theory of Modernity* (New York: Columbia University Press).

Rose, Todd. 2016. *The End of Average: How We Succeed in a World That Values Sameness* (New York: HarperOne).

Rovelli, Carlo. 2016. *Reality is Not What it Seems: the journey to quantum gravity* (London: Allen Lane, UK).

Rowan Tiddy. 2013. *The Little Book of Mindfulness* (London: Quadrille).

Rowling, J. K., Thorne, J and Tiffany, J. 2016. *Harry Potter and the Cursed Child* (New York: Arthur A. Levine Books).

Rubenstein Mary-Jane. 2012. "Cosmic Singularities: On the Nothing and the Sovereign." *Journal of the American Academy of Religion* 80.2: 485–517.

Rudolf, Anthony. 1999. *The Arithmetic of Memory* (London: Bellew Publishing).

Sabbagh, Karl. 2010. *Remembering our Childhood: how memory betrays us* (Oxford: Oxford University Press).

Sacks, Oliver, 2013. "Speak, Memory," *New York Review of Books*, 21 February.

Schopenhauer, Arthur. 2012. *Essays of Schopenhauer*. Trans. Mrs Rudol Dircks (Adelaide: University of Adelaide).

Schüll, Natasha Dow. 2012. *Addiction by Design: Machine Gambling in Las Vegas* (Princeton: Princeton University Press).

Segal, Suzanne. 1996. *Collision with the Infinite* (San Diego: Blue Dove Press).

Sieborg, Steve. 2010. *How Rich People Think* (White Fox Cove, OH: London House Press).

Singer, Richard, A. 2011. *Now: Embracing the Present Moment* (Airesford, Hants: O-Books).

Smyth Damien, Mark. 2012. *Do Nothing: Stop Looking, Start Living* (Corby, UK: 3P Publishing, UK).

Storr, Will. 2013. *The Heretics: Adventures with the Enemies of Science* (London: Picador).

Sunim Haemin, 2017. *The Things You Can See Only When You Slow Down* (New York: Penguin Books); Tra edition.

Taleb, Nassim Nicholas. 2010. *The Black Swan* (New York: Penguin).

Tammet, Daniel. 2012. *Thinking in Numbers: How Maths Illuminates Our Lives* (London: Hodder & Stoughton).

Taylor, Steven. 2008. *Making Time: Why Time Seems to Pass at Different Speeds and How to Control it* (London: Icon Books).

Tetlock, Philip E. and Gardner, Dan. 2015. *Superforecasting: The Art and Science of Prediction* (New York: Broadway Books).

Thurber, James. 1950. *The 13 Clocks (Children's Collection)* (New York: Simon & Schuster).

Tolle, Eckhart. 1999. *Practising the Power of Now* (Eckhart Tolle, CA: New World Library).

----- 2004. *The Power of Now: A Guide to Spiritual Enlightenment* (Eckhart Tolle, CA: New World Library).

Trevor-Roper, Hugh. 2009. *The Invention of Scotland* (New Haven, CT: Yale University Press).

Tulgan, Bruce. 2009. *Not everyone Gets a Trophy* (New York: Jossey-Bass).

Unger, Roberto Mangabeira. 2014. *The Religion of the Future* (Cambridge, MA: Harvard University Press).

Van de Ven, N., Van Rijswijk, L. and Roy, M. M. 2011. "The return-trip effect: Why the return-trip often seems to take less time." *Psychonomic Bulletin & Review*, 18, pp. 827–832.

Vonnegut, Kurt. 1998. *Timequake* (New York: Vintage).

Waller, Robert James. 1992. *The Bridges of Madison County* (Warner Bros).

Warhol, Andy. 1977. *The Philosophy of Andy Warhol (From A to B and Back Again)* (Eugene, Oregon: Harvest).

Watts, Alan. 1953. *The Wisdom of Insecurity* (New York: Vintage Books).

Watts, Steven. 2014. *Self-help Messiah: Dale Carnegie and Success in Modern America* (New York: Other Press).

Wheeler, John. 1990. "Information, physics, quantum: The search for links." In *Zurek, Wojciech Hubert. Complexity, Entropy, and the Physics of Information* (Redwood City, California: Addison-Wesley).

Wiesentier, Leon. 2013. "Idealism and Blindness: of flaking paint and blemishes." *New Republic*, June 10, 2013.

Williams, Mark and Danny Penman. 2011. *Mindfulness: An Eight-Week Plan for Finding Peace in a Frantic World* (New York: Rodale).

Wright, Robert. 2010. *The Evolution Of God* (New York: Back Bay Books).

Yalom, Irvin, D. 1989. *Love's Executioner: and Other Tales of Psychotherapy* (New York: Basic Books).

Yorke, John. 2013. *Into The Woods: A Five-Act Journey Into Story* (New York: Penguin Books).

Index

BOOKS

Iff Books

ACADEMIC AND SPECIALIST

Iff Books publishes non-fiction. It aims to work with authors and titles that augment our understanding of the human condition, society and civilisation, and the world or universe in which we live.

If you have enjoyed this book, why not tell other readers by posting a review on your preferred book site.

Recent bestsellers from Iff Books are:

Why Materialism Is Baloney

How True Skeptics Know There is no Death and Fathom Answers to Life, the Universe, and Everything

Bernardo Kastrup

A hard-nosed, logical, and skeptic non-materialist metaphysics, according to which the body is in mind, not mind in the body.

Paperback: 978-1-78279-362-5 ebook: 978-1-78279-361-8

The Fall

Steve Taylor

The Fall discusses human achievement versus the issues of war, patriarchy and social inequality.

Paperback: 978-1-90504-720-8 ebook: 978-184694-633-2

Brief Peeks Beyond

Critical Essays on Metaphysics, Neuroscience, Free Will, Skepticism and Culture

Bernardo Kastrup

An incisive, original, compelling alternative to current mainstream cultural views and assumptions.

Paperback: 978-1-78535-018-4 ebook: 978-1-78535-019-1

Framespotting

Changing How You Look at Things Changes How You See Them

Laurence & Alison Matthews

A punchy, upbeat guide to framespotting. Spot deceptions and hidden assumptions; swap growth for growing up. See and be free.

Paperback: 978-1-78279-689-3 ebook: 978-1-78279-822-4

Is There an Afterlife?

David Fontana

Is there an Afterlife? If so what is it like? How do Western ideas of the afterlife compare with Eastern? David Fontana presents the historical and contemporary evidence for survival of physical death.

Paperback: 978-1-90381-690-5

Nothing Matters

A Book About Nothing

Ronald Green

Thinking about Nothing opens the world to everything by illuminating new angles to old problems and stimulating new ways of thinking.

Paperback: 978-1-84694-707-0 ebook: 978-1-78099-016-3

Panpsychism

The Philosophy of the Sensuous Cosmos

Peter Ells

Are free will and mind chimeras? This book, anti-materialistic but respecting science, answers: No! Mind is foundational to all existence.

Paperback: 978-1-84694-505-2 ebook: 978-1-78099-018-7

Punk Science

Inside the Mind of God

Manjir Samanta-Laughton

Many have experienced unexplainable phenomena; God, psychic abilities, extraordinary healing and angelic encounters. Can cutting-edge science actually explain phenomena previously thought of as 'paranormal'?

Paperback: 978-1-90504-793-2

The Vagabond Spirit of Poetry

Edward Clarke

Spend time with the wisest poets of the modern age and of the past, and let Edward Clarke remind you of the importance of poetry in our industrialized world.

Paperback: 978-1-78279-370-0 ebook: 978-1-78279-369-4

Readers of ebooks can buy or view any of these bestsellers by clicking on the live link in the title. Most titles are published in paperback and as an ebook. Paperbacks are available in traditional bookshops. Both print and ebook formats are available online.

Find more titles and sign up to our readers' newsletter at http://www.johnhuntpublishing.com/non-fiction

Follow us on Facebook at https://www.facebook.com/JHPNonFiction and Twitter at https://twitter.com/JHPNonFiction